Making Complex Decisions toward Revamping Supply Chains amid COVID-19 Outbreak

Making Complex Decisions toward Revamping Supply Chains amid COVID-19 Outbreak

Edited by
Dinesh Kumar and Kanika Prasad

CRC Press
Taylor & Francis Group
Boca Raton London

CRC Press is an imprint of the
Taylor & Francis Group, an **informa** business

First edition published 2022
by CRC Press
6000 Broken Sound Parkway NW, Suite 300, Boca Raton, FL 33487–2742

and by CRC Press
4 Park Square, Milton Park, Abingdon, Oxon, OX14 4RN

CRC Press is an imprint of Taylor & Francis Group, LLC

ISBN: 978-0-367-71265-5 (hbk)
ISBN: 978-0-367-71266-2 (pbk)
ISBN: 978-1-003-15008-4 (ebk)

DOI: 10.1201/9781003150084

Typeset in Times
by Apex CoVantage, LLC

To my father Shri Suresh Pal, my mother Smt. Sudesh Devi, my beloved wife Sarita, sons Anay, Rishit, and all the contributors of this book.

Dinesh Kumar

To my family members, mentors, and friends.

Kanika Prasad

Contents

Preface

The COVID-19 disease outbreak was declared a global health emergency by the World Health Organization (WHO) in January 2020. The disease spread worldwide to become a global pandemic in a few weeks and posed an unprecedented public health and safety challenge, resulting in enormous loss of human lives and severe health issues for many infected with the virus. In addition, the devastating impact of the pandemic has led to unprecedented social and economic disruptions, thus changing the way global economic activities take place. To contain the virus and manage the situation effectively, governments worldwide declared emergencies, imposed lockdowns, restricted travel, raised taxes, increased surveillance, and curtailed rights. The effects of the worldwide lockdown were devastating to the global economy. The lockdown disrupted both supply and demand ends. The constrained movement of goods and services led to raw materials scarcity resulting in commodity price inflation, restricted migration of skilled workers, warehouse losses, and often partial or complete loss of demand. This disrupted the global supply chain in the short to medium term, and on a macro level, this event has sowed the seeds of major geopolitical and trade order realignment in the long term.

This book aims to explore the supply chain challenges that appeared due to COVID-19 disruption and develop resilient supply chain models to counter such a disruption. Structural mapping of the parameters leading to socioeconomic challenges during the pandemic is presented. Further inventory management, especially perishables, is a significant concern since it drives the entire supply chain. The same becomes crucial in the pandemic restrictions as the inventory deteriorates with respect to time. The appropriate inventory policies for perishables in lockdown situations under disrupted production and demand are presented with numerical examples and sensitivity analysis.

Achieving sustainability amid such a disruptive scenario is a necessity. It is a significant challenge for organizations and is one of the focus areas of this book. Sustainability in a supply chain can only be achieved through strategic, tactical, and operational level decisions, especially in a disruptive environment like a pandemic. Therefore, various sustainable strategies under the effect of the COVID-19 pandemic are described in this book. Additionally, carbon emissions through multiple stages of a supply chain is a major concern around the globe. Developed and developing nations are striving to reduce such emissions in several ways, including carbon credit and investment in energy-efficient or green technologies. Given this, a two-warehouse sustainable inventory model for perishable items has been presented considering the demand disruption due to COVID-19 and carbon emissions taxation. Furthermore, this book also covers critical issues related to the pandemic, including supplier selection, achieving supply chain resiliency through Industry 4.0, and analyzing corporate social responsibilities.

Further, this book is organized as follows:

Chapter 1 – This chapter analyzes how the COVID-19 crisis affects the socio-economic situation and panic-buying behaviour of the public at the local level and identifies immediate government interventions to respond to the problem. This chapter proposes a framework to assess the actual situation and decide upon the management measures in the Indian context.

Chapter 2 – A production-inventory model for perishable items based on a manufacturing facility and a warehouse is presented in this chapter. The deterministic model has been developed for two different scenarios depending on the lockdown duration: shorter and extended periods. In addition, a constant deterioration rate and a selling price-dependent demand rate with production rate depending upon demand have been considered. The demand and production in this model decrease as the lockdown is imposed and increase when relaxed.

Chapter 3 – There is an intense need to retain the sustainability goals of supply chain management in this difficult pandemic situation. Maintaining the smoothness of the operations during disruptions is also important. The study proposes a human-centric decision framework to integrate supply chain resilience (SCR) and sustainable supply chain (SSC) strategies to achieve the supply chain objectives.

Chapter 4 – This chapter identifies the strategies to develop a sustainable supply chain amid the pandemic like scenarios. Various strategies are determined from the literature review, and survey research is carried out using the questionnaire method. The sectors identified for the survey are the automobile, electronics, retail, textile, and chemical industries.

Chapter 5 – The chapter highlights the role of the circular economy (CE) to reduce the perils caused by pandemic outbreaks by using its evolutions in resource reutilization and regeneration. Here, nine prominent business sectors are identified. The effect of the outbreak is analyzed, and an attempt made to provide the measures for mitigating these perils through the application of circular economy business models (CEBM). In this research, the proposed measures to the identified business sector would enable its firms to compete with the present miserable condition due to the outbreak and strengthen them for future challenges.

Chapter 6 – In this chapter, two scenarios of a finite period complete lockdown in the proposed model are considered in two warehouse environments for deteriorating items to minimize the total system cost. In the first scenario, it is assumed that the implementation of lockdown and relief in stringent containment measures are given when the items are being supplied from a rented warehouse. In contrast, the second scenario assumes an extensive lockdown period that lasts when the items are being dispatched from the owned warehouse.

Chapter 7 – Selection of suppliers amid the COVID-19 pandemic is a critical step in the supply chain and requires making complex decisions involving various factors, often conflicting in nature. A software prototype developed in this chapter provides tools for the industry to cope with such scenarios and take appropriate decisions to come out unscathed.

Chapter 8 – Adoption of new technologies such as Industry 4.0, which encompasses many technologies including the Internet of Things (IoT), artificial intelligence (AI), machine learning (ML), big data, augmented reality (AR), virtual reality (VR), and cloud computing, will gather pace in manufacturing and its supply chain as the economy emerges out of the pandemic phase. IoT implementation will help remote monitoring of manufacturing operations by integrating machines, equipment, and workforce and improving plant safety. Adopting AI and ML will help bring further efficiency in planning, monitoring, and executing logistics operations. Cloud technology implementation can quicken and streamline the development of digital solutions. For example, virtual assistants or bots can be employed for aftermarket sales service and at the supply chain command center.

Chapter 9 – Corporate social responsibility (CSR) activities relevant to times of distress can effectively satisfy various stakeholders and society at large to create a positive impact. Tools for the selection of CSR activities with maximum impact are discussed here. This chapter demonstrates the application of an integrated QFD-TOPSIS technique for selecting the best CSR project of an Indian enterprise to fulfill various objectives of assorted stakeholders of its value chain. In addition to analyzing the CSR programs based on identified technical requirements, the proposed technique provides improvement areas for projects that show huge deviation, thus helping the management of a company to improve the effectiveness of plans, policies, and strategies.

Acknowledgements

First and foremost, the editors take this opportunity to thank each of the authors for their contribution to the book. Our sincere gratitude to them for contributing their valuable time, resources, and expertise to make this book possible. Second, the editors are also grateful to all the reviewers for taking time out from their busy schedules and providing valuable suggestions and observations, resulting in improved quality, rationality, and content presentation of the chapters.

The editors owe a deep debt of gratitude to CRC Press/Taylor & Francis Group for providing the opportunity to publish this book. Their constant guidance and motivation made it possible for us to lay the foundation of this book.

The editors would like to thank all the readers for their trust and sincerely hope that this book will help them in their future endeavorus.

We would fail in our duties if we did not express our heartfelt thanks to our family members who have been a constant source of love, support, concern, strength, and caring in the difficult prevalent situation of the global pandemic.

The editors gratefully acknowledge the support of all the people who directly or indirectly rendered their assistance, constant encouragement, and help in the production of this book.

Finally, praises and thanks to God for His blessings to complete the book successfully. None of this would have been possible without the blessings of the Almighty.

Editors

Dinesh Kumar, PhD, is an Assistant Professor in the Department of Production and Industrial Engineering, National Institute of Technology, Jamshedpur, Jharkhand, India. He completed his PhD from IIT Roorkee in 2016 in supply chain management. He has authored several papers in reputed journals. His areas of expertise are perishable inventory planning, operations management, supply chain management, and system dynamics. Dr. Kumar has guided a number of master's theses in his research acumen. He has authored various book chapters in several international editions. He is the co-editor of the book entitled *Agri-Food 4.0: Innovations, Challenges and Strategies* (Emerald Publishing). He is the reviewer of various reputed international journals, including *Operations Management Research*, *International Journal of Logistics Research and Applications*, and *Computers and Industrial Engineering*. He is a life member of the Operations Research Society of India (ORSI).

Kanika Prasad, PhD, is an Assistant Professor in the Production and Industrial Engineering Department of the National Institute of Technology, Jamshedpur. She graduated in mechanical engineering from Sikkim Manipal Institute of Technology, Sikkim, in 2010 and completed her master of engineering in production engineering from Jadavpur University, Kolkata, in 2013, earning distinction in both degrees. She earned a prestigious DST INSPIRE fellowship to pursue her doctoral research from Jadavpur University, where she earned her PhD in 2016. Dr. Prasad worked with Sikkim Manipal Institute of Technology as Assistant Professor for close to a year and taught various undergraduate and post-graduate subjects, such as operations research, quality control assurance and reliability, manufacturing planning and control, and manufacturing process. She published several papers in peer-reviewed international journals. With many publications in journals and conferences of international repute so early in her academic career, much is expected of her research career. Dr. Prasad specializes in the application of quality function deployment technique and multi-criteria decision-making to develop expert systems for selection and design problems in manufacturing. Of late, sustainability and waste management domain interests her and she is working in that area. She has guided several UG and PG dissertations. She is also a regular reviewer of several journals of international repute.

Contributors

Somesh Agarwal
Department of Industrial and Production Engineering
Dr B R Ambedkar National Institute of Technology
Jalandhar, Punjab, India

Manish Mohan Baral
Department of Operations
GITAM Institute of Management, GITAM (Deemed to be University)
Visakhapatnam, Andhra Pradesh, India

Neeraj Bhanot
CSIR – National Physical Laboratory
New Delhi, India

Rishabh Bhanot
Oral and Maxillofacial Surgery
Jyoti Kendra General Hospital
Ludhiana, Punjab, India

Leopoldo Eduardo Cárdenas-Barrón
Tecnologico de Monterrey, School of Engineering and Sciences
Monterrey, NL, Mexico

Ratnesh Chaturvedi
Department of Finance Management
Xavier Institute of Social Service
Ranchi, Jharkhand, India

Venkataiah Chittipaka
Operations Management
Indira Gandhi National Open University
Delhi, India

Rishi Dwivedi
Department of Finance Management
Xavier Institute of Social Service
Ranchi, Jharkhand, India

R.K. Garg
Department of Industrial and Production Engineering
Dr B R Ambedkar National Institute of Technology
Jalandhar, Punjab, India

Hritik Kabra
Dr B R Ambedkar National Institute of Technology
Jalandhar, Punjab, India

Harshit Katurka
Department of Production and Industrial Engineering
National Institute of Technology Jamshedpur
Jamshedpur, Jharkhand, India

Amanpreet Kaur
Royal Bank of Scotland
New Delhi, India

Amaresh Kumar
Department of Production and Industrial Engineering
National Institute of Technology Jamshedpur
Jamshedpur, Jharkhand, India

Dinesh Kumar
Department of Production and Industrial Engineering
National Institute of Technology Jamshedpur
Jamshedpur, Jharkhand, India

Bijaya K. Mangaraj
Production, Operations and Decision Sciences
XLRI Xavier School of Management
Jamshedpur, Jharkhand, India

Rahul S. Mor
National Institute of Food Technology Entrepreneurship and Management
Sonepat, India

Arup Mukherjee
Department of Finance Management
Xavier Institute of Social Service
Ranchi, Jharkhand, India

Subhodeep Mukherjee
Department of Operations
GITAM Institute of Management, GITAM (Deemed to be University)
Visakhapatnam, Andhra Pradesh, India

Vikash Murmu
Department of Production and Industrial Engineering
National Institute of Technology Jamshedpur
Jamshedpur, Jharkhand, India

Ankit Nayan
SynaptoCrats
Ludhiana, Punjab, India

Mohammad Faisal Noor
Department of Production and Industrial Engineering
National Institute of Technology Jamshedpur
Jamshedpur, Jharkhand, India

Surya Kant Pal
University School of Business – ICP
Chandigarh University
Mohali, Punjab, India

Kanika Prasad
Department of Production and Industrial Engineering
National institute of Technology Jamshedpur
Jamshedpur, Jharkhand, India

Samidha Prasad
HDFC Bank Ltd.
Patna, Bihar, India

Piyush Rai
Department of Computer Science & Engineering, I.E.T
Dr. R.M.L. Avadh University
Ayodhya, Uttar Pradesh, India

Ranveer Singh Rana
Department of Production and Industrial Engineering
National Institute of Technology Jamshedpur
Jamshedpur, Jharkhand, India

Biswajit Sarkar
Department of Industrial Engineering
Yonsei University
Seoul, South Korea

Varun Sharma
Production, Operations and Decision Sciences
XLRI Xavier School of Management
Jamshedpur, Jharkhand, India

Nikita Sinha
Department of Production and Industrial Engineering
National Institute of Technology Jamshedpur
Jamshedpur, Jharkhand, India

Smita
Department of Mathematics
Government Polytechnic
Patna, Bihar, India

Amar Eron Tigga
Department of Marketing Management
Xavier Institute of Social Service
Ranchi, Jharkhand, India

Mohit Tyagi
Department of Industrial and Production Engineering
Dr B R Ambedkar National Institute of Technology
Jalandhar, Punjab, India

1 Mapping the Socio-Economic Impact of COVID-19 in Indian Context

The Way Forward

Neeraj Bhanot, Rahul S. Mor, Hritik Kabra, Ankit Nayan, and Rishabh Bhanot

CONTENTS

1.1 INTRODUCTION

Humanity has been ravaged by plagues and pandemics, often changing the framework of the functioning of the globe. The monstrous and deadly family of human pathogens causes various types of diseases, including the common cold and more painful illnesses that result in the untimely death of the infected. Due to the crown-shaped protrusions around itself, the virus has been named coronavirus. The effects of COVID-19 were first discovered in Wuhan, China, in December 2019, and now nearly the whole world is in the grips of the pandemic (WHO, 2020).

DOI: 10.1201/9781003150084-1

According to the WHO, since the symptoms are curable, hence, the treatment is based on the severity of the patient (WHO, 2020). In India, where a significant part of the revenue is generated by tourism and continued supply chain processes, the country will be under the impression of decline amid prolonged quarantine, thereby weakening the global economy or even causing a recession. Researchers predicted that the pandemic would undoubtedly cause the gross domestic product (GDP) to reach 0% in the year 2020 (Mishra, 2020).

Globally, governments face an alarming hike in the number of patients and have taken some bold steps to protect their citizens. India, too, took a bold step by implementing a harsh lockdown throughout the nation for 21 days, which can be considered one of the most extensive lockdown periods ever held (Gupta & Madgavkar, 2020). The halted economic development, rise in unemployment, and increasing severity of the crisis have made it essential to bring out some evolutionary measures to restore the conditions.

Researchers, via harmonized technology, mapped out measures and their implementation by voluntary and enforceable plans, giving a ray of hope for countries worst affected like Italy (Bruin et al., 2020). Intermittently, Asian countries too took immediate action after witnessing the situation in Italy, and nations are putting in rigorous efforts to resume their previous life in the country (Waris et al., 2020).

The battle is already half won if one is confident enough and free from all fears that signify the need to take the citizens' mental situation as a matter of concern (Roy et al., 2020). Moreover, researchers shared several operational guidelines and strategies to be instantly implemented for minimizing the spread of COVID-19 as well as helping to maintain all the clinical and educational needs of hospitals (Goh et al., 2020).

Based on identified literature, it can be seen that there is minimal discussion about the working of different sections of society in fighting against the crisis. Thus, the purpose of this study is to scrutinize the synergy among the people of the nation in reducing the spread of COVID-19, which in turn will give out some fruitful results which no doubt can be analyzed to counter the situation of the pandemic in any country. The methodology used signifies a completely result-oriented approach to describe the conditions in a particular area. Dividing society into four quadrants and accordingly defining some indicators to quantify their effectiveness has been done. Moreover, using these indicators, different parts of India are listed as critical places by health advisories.

Objective-oriented research done in this study will not only help to improve the condition of the crisis in different parts of the nation, it will also sum up the efforts of the four quadrants of society. The results are crucial for completely turning the situation at worst affected parts upside down and highlighting the loopholes in the districts acclaimed as models that could help them in completely uprooting the traces of COVID-19.

1.2 FOUR QUADRANTS OF SOCIETY

The national emblem of the Republic of India is an adaptation of the Lion Capital of Ashoka, which features four lions symbolizing power, courage, confidence, and pride. Similarly, the people of India working diligently against COVID-19 are termed as COVID-19-warriors/COVID-19-fighters by everyone and comprise police officers, doctors, nurses, sanitation service members, delivery people, and more. So, to analyze their coordination, an exemplary display of grit and determination, these people have been divided into four quadrants as follows:

1. *Administration/Policymakers:* The first quadrant of COVID-19 warriors comprises administration and policymakers, who are working day and night to prevent the spread of COVID-19. Be it imposing a severe lockdown, cluster mapping, identifying hotspots, or testing and appealing, the fighters have done all to protect their citizens. Additionally, the individuals participating in numerous hackathons and competitions by the government to generate ideas to combat COVID-19 and the manufacturers working tirelessly in manufacturing daily needs are undoubtedly an aid to the first quadrant and made their efforts fruitful.
2. *Population at Home:* Second quadrant warriors are the population at home consistently following the instructions of the first quadrant for reducing the impact of COVID-19 and are undoubtedly a major support. Whether it is cancelling all their once in a lifetime gatherings or sacrificing their time for the betterment of the nation, they deserve a bow.
3. *Essential Service Providers:* Third quadrant COVID-19 warriors are similar to the fourth lion of our emblem, which is hidden, but without its support the efforts of the other three are incomplete. Essential service providers like sanitation workers, non-governmental organizations (NGOs), delivery people, vendors of daily needs, and media persons who provide updates made the sacrifice of the population at home worthy. They didn't make even a single person step out of the home by fulfilling every need in no time.
4. *Healthcare Professionals*: Last but not least, the fourth quadrant of COVID-19 warriors are none other than our healthcare professionals, doctors, nurses, ward boys, compounders, cleanliness servicemen, and so on who are working in firefighting mode against the deadly virus. Efforts by the other three quadrants act as a booster for them to work diligently without being under any other stress.

However, the contribution of all four quadrants has been visualized based on the selection of a few critical performance indicators and is explained in the next section.

1.3 METHODOLOGY

The methodology of the research work was based on scrutinizing the data available regarding COVID-19 on various platforms such as newspaper articles, web sources, and statistics released by the healthcare bodies alongside numerous briefing sessions of state officials. The data has been analyzed to be used as a foreseeing tool to improve future proceedings and measures for the containment of COVID-19 in the country. The database for different indicators has been updated to April 17–18, 2020, and thus has been set as the period of research.

The data was mainly collected in two phases, first for the current scenario of states and then for the districts. In the first phase, the current position of various states in measures of COVID-19 infectivity was assessed based on an article published in *The Hindu* (The Hindu Data Team, 2020), which had arranged the states according to the seriousness of COVID-19 pandemic to date. Therefore, selecting the top ten critical states was done to analyze the severity of the crisis in those places. However, data in phase two was collected by obtaining data of those districts where the central government was applauded for containing COVID-19 to a great extent, namely Bhilwara (FE Online, 2020; Sharma, 2020), Kasaragod (Jayarajan, 2020b; Koshy, 2020), and Agra (FE Online, 2020). Similarly, the top three worst cases were chosen based on the risk of community spread as the number of patients found in these districts was seemingly much higher than others (Bhatia & Devulapalli, 2020), namely Mumbai (ET Bureau, 2020a; Qureshi, 2020b), Indore (Mekaad & Singh, 2020; PTI, 2020d), and Jaipur (Bohra, 2020). The details on various indicators have been provided as follows:

1. Tested Population: Refers to the population tested in regards to finding the possibility of COVID-19.
2. Test per Million People: Refers to performing several tests being done per million people. While data for the state was available (The Hindu Data Team, 2020), in the case of districts, the number was calculated based on the following formula:

   $$\text{Tests per Million People} = (\text{Tested Population}/\text{Total Population}) \times 10^6$$

   Wherein; Total Population for District was referred from the World Population Review (World Population Review, 2020).
3. Cases Confirmed: Refers to the number of patients who tested positive for COVID-19, signifying the amount of spread.
4. Cases Recovered: Refers to the patients successfully cured out of positive patients, explaining medical outputs.
5. Deaths: Refers to the number of patients losing their lives due to COVID-19 infection.
6. Corona Case Positivity (%): This refers to the spread of COVID-19 in the places, signifying the extent of the spread of COVID-19 in suspicious persons and is calculated as follows:

   $$\text{Corona Case Positivity} = (\text{Confirmed Cases}/\text{Tested Population}) \times 100$$

7. Percentage of Active Cases (%): Refers to the percentage of patients still undergoing treatment of COVID-19, which signifies the current scenario of the spread in a particular area and is calculated as follows:

   $$\%\text{Active Cases} = [\{\text{Cases Confirmed} - (\text{Cases Recovered} + \text{Deaths})\}/\text{Cases Confirmed}] \times 100$$

8. Rate of Recovery (%): Defined as how efficiently healthcare professionals are curing the patients, signifying healthcare effectiveness in curing patients and is calculated as follows:

   $$\text{Rate of Recovery} = (\text{Cases Recovered}/\text{Cases Confirmed}) \times 100$$

9. Mortality Rate (%): Compares the number of patients who died while in treatment to that of patients who tested positive, which signifies the criticality of the patients and is calculated as follows:

$$\text{Mortality Rate} = (\text{Deaths/Case Confirmed}) \times 100$$

Thus, based on these indicators, results for selected states and districts have been suitably discussed in the next section.

1.4 RESULTS AND DISCUSSION

The results of states and districts for indicators have been calculated and tabulated in the Appendix. Based on analysis, the discussion in two stages has been presented as follows:

1.4.1 Results Based on State Analysis

Figure 1.1 presents plots for different combinations of indicators comparing the scenarios by taking two indicators at a time.

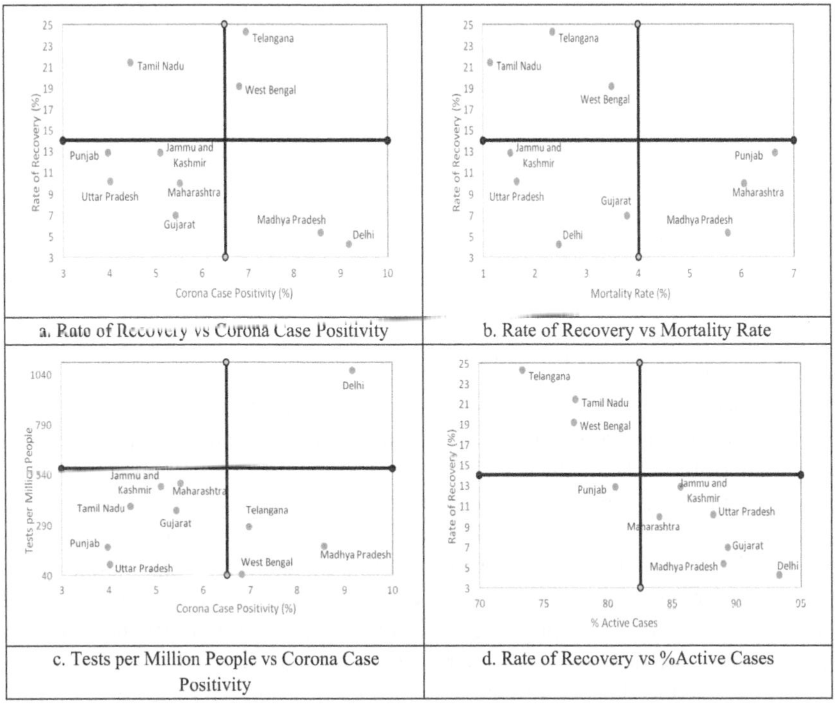

a. Rate of Recovery vs Corona Case Positivity

b. Rate of Recovery vs Mortality Rate

c. Tests per Million People vs Corona Case Positivity

d. Rate of Recovery vs %Active Cases

FIGURE 1.1 State-based plots for different indicators.

Plotting rate of recovery against corona case positivity gave insights, from the tragic situation at Delhi with the least rate of recovery of merely 4.21% to the highest corona case positivity of 9.17% among all states, which is no doubt a matter of concern. The situation at the capital turned into a crisis when it was found that a congregation resulted in many COVID-19 patients (Roy, 2020). Conversely, Telangana's response to curing outpatients is admirable with over 24.28% rate. Meanwhile, Tamil Nadu, with its unique countermeasures, had both lower corona case positivity of 4.45% and a high rate of recovery of 21.39%.

Tamil Nadu stood firm like a mountain against the death and obtained the lowest mortality rate of 1.13% among all other states. Madhya Pradesh was in a dwindling phase with an almost similar mortality rate to that of the recovery rate with 5% each, which was not a good outcome because of the generation of two hotspots in Indore and Bhopal with over 80% cases (Dwary & Nair, 2020).

West Bengal state was no different in being affected with COVID-19 like the rest of the states in India. Still, the state had the lowest testing per million done of merely 42.7, which was undoubtedly very low considering its high corona case positivity of 6.81% (Ghosh, 2020b).

While Delhi is the most badly affected among all with 9.17% corona case positivity, they did an admirable job in reaching out, with the maximum tests per million among all with over 1060 tests.

Gujarat's land of the white desert showed an alarming situation in the state, with more than 89.30% active cases to a significantly low 6.91% recovery rate. According to the administration, the reason behind the spike in the number of patients in Ahmedabad and other parts of Gujarat was extensive testing, which identified hotspots accurately within the state (PTI, 2020b). With over 93.31% active cases to a 4.21% recovery rate, the country capital seemingly approached a situation similar to community spread. Following the trend, Madhya Pradesh and Uttar Pradesh had a stressful above 85% active cases, which in no time will become a more and more tense situation.

In some states like Tamil Nadu and Telangana, rigorous administration efforts had proved to be fruitful in achieving a high recovery rate (both above 20%) and significantly lower active cases (both below 80%), which itself made them an example for other states (for COVID-19 statistics, please refer to Appendix – Table 1.A).

1.4.1.1 Issues to Be Addressed for States

The biggest question the whole country is seeking an answer for is why the congregation at Delhi happened amid this fragile situation that blasted into hotspots for almost every state of India. Undoubtedly, the administration did every negotiation with the preachers, but it was a blind "no" from the other side, and the rest is pandemic.

Another question that is generating debate is the use of rapid testing techniques for COVID-19 testing. Some say it gives out exact numbers, while others suggest stopping it for reducing havoc and panic among people. Surprisingly, states like Telangana and Tamil Nadu played great innings against COVID-19 and can sooner reach a better situation, whose measures of cluster mapping and strict surveillance can be suggested to the states fighting against COVID-19 that have not yet reached a stable level.

This pandemic could be treated as a blessing in disguise by analyzing the need for healthcare units in the states alongside how optimistic approaches can curb such a crisis.

The research done in the current circumstances can help assess and prioritize the resource allocation efficiently. It suggests to the authorities to maintain adequate availability of medical resources at Delhi, Madhya Pradesh, and Gujarat, which have a very high percentage of active cases leading to an increase in the mortality rate, so that the rate does not rise due to lack of ventilators and personal protective equipment. It can be regarded as foreseeing the damage and reducing its impact to a maximum extent.

1.4.2 Results Based on District Analysis

Districts have been categorized into two types: best-case scenarios comprising Bhilwara, Agra, and Kasaragod, and worst-case scenarios comprising Mumbai, Indore, and Jaipur. Figure 1.2 similarly presents plots for different combinations of indicators for districts to compare the scenarios.

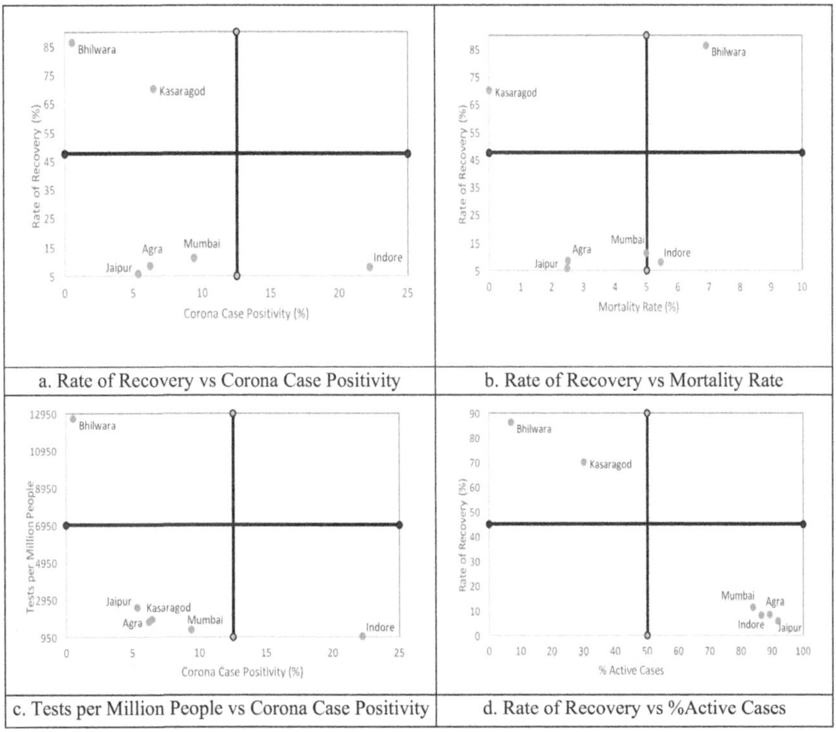

FIGURE 1.2 District based plots for different indicators.

1.4.2.1 Best-Case Scenarios

- **Bhilwara Model**
  - Amid all the chaos, Bhilwara, a small district of Rajasthan, became the most significant talk of the nation when it was found to contain more than 5000 people (Sharma, 2020) suspected of being under the effect of the deadly virus. The situation would have resulted in the replication of Bhilwara into the situation in Italy in no time. Still, the devoted, brisk, and tireless efforts of all four quadrants of warriors freed the district from the shackles of COVID-19.

    With the aid of police officers, the administration and policymakers imposed an instant and strict curfew, thereby restricting the extension of the affected people. Equipped with the required information, they started cluster mapping alongside door-to-door screening, therefore paving the way for clearly dividing the city into various zones that included self-quarantine high-risk zones, moderate zones, and less severe zones (Bhatt, 2020).

    Over 12,651 tests per million were done to identify the infected people in the district. Bearing out favorable results of only 0.5% being positive out of the total tested, the ruthless lockdown led to almost no spreading of the virus. Henceforth, the social transmission of the virus did not happen, leading to a halt in increasing the number of patients.

    The administration was well aware that people cannot withstand an extended lockdown without being equipped with daily needs. Nonetheless, the essential service providers, unsung warriors of the society, stood up. They contributed to rectifying this stressful situation by relentlessly doing their services without worrying about their own health.

    Their selfless efforts proved fruitful as 25 out of 29 patients were recovered with a whopping efficiency of 86%. Doctors are fighting this fragile situation in firefighting mode by maintaining a medical emergency of 6.8% to cure the patients infected at the earliest. Unfortunately, the mortality rate was observed to be 6.89%, mainly due to the previous poor medical history of the patients (for COVID-19 statistics, please refer to Appendix – Table 1.B, S. No. 4).

- **Kasaragod Model**
  - A place usually known for its eternal beauty comprising the seashore, natural habitats, and historical architecture soon closed its doors when it was found to be affected with COVID-19. Moreover, many of the Kasaragod district's population travel abroad, which resulted in two waves of infections, first in February via a China-returned student and the second in mid-March from travelers coming from Middle East countries (Jayarajan, 2020a). So, subsequently, the COVID-19 positivity rose to 6.42% in the district in no time.

    However, given the tremendous effort by four quadrants of society alongside learning from the state managing the earlier fallout from the

SARS and Nipah outbreaks, Kasaragod successfully implemented a model to avert the looming infection.

The district administration and policymakers used geospatial tracking through drones to put people at home to ensure the successful implementation of Section-144. Besides, they urged people who had traveled from other countries to be in self-quarantine and ran an effective campaign named *#BreakTheChain* to create awareness for social distancing (Jayarajan, 2020a). Conversely, people staying at their homes proved to be a helping hand for the administration by flattening the curve and bringing down the positivity rate, with a 68.45% recovery rate so far, one of the highest in the country.

Nonetheless, the third quadrant of the society ensured that the poor and needy got free food and regular health checkups through "Care for Kasaragod" and "Jana Jagratha Samithi" at the ward level for free food distribution (Jayarajan, 2020a).

Shelter homes and community kitchens were also started to achieve a complete lockdown in the district. Astonishingly, with the help of Accredited Social Health Activist (ASHA) workers and health inspectors, 17,300 contacts of positive cases were not only traced and quarantined but were also tracked day and night during the entire incubation period by healthcare professionals (Ghosh, 2020a; HT Correspondent, 2020).

To further control the situation, a testing rate at 1882.14 tests per million were undertaken wherein whenever the tested people reported any symptoms, they were presumed to be COVID-19 positive and were instantly transported to isolation centers.

Undoubtedly after such strict and aggressive measures, almost all contacts were identified and followed up. This resulted in districts having a 0% mortality rate, which was a record in itself. Furthermore, healthcare professionals' organized and streamlined efforts had recovered many patients with an efficiency of 70.05%, which kept the situation intact and controlled (for COVID-19 statistics, please refer to Appendix – Table 1.B, S. No. 6).

- **Agra Model**
  - With the increasing number of suspected COVID cases, Agra appeared as one of the worst affected COVID-19 districts in India; around 4000 people were suspected in the initial days that quickly reached 6.19% in Taj, which was unquestionably the signs of an alarming situation (Qayam, 2020). However, due to the tremendous effort of all four quadrants of society, Agra successfully implemented a model to contain the spread of COVID-19 that emerged as a good cluster containment strategy.

    The administration, along with the police, implemented complete lockdown and sealed borders. Also, the Agra Smart City project team (ICCC) made this lockdown an effective one by developing a mobile application, thus alerting the crowded places (SCC, 2020).

By the administration's quick decision, a team of over 2000 nursing staff, hand in hand with 3,000 ASHA workers, surveyed door to door and figured out most suspected cases (Chaturvedi, 2020). The team covered 160,000 houses, which resulted in 1760.43 tests per million in the district. Which, in turn, resulted in finding the epicenter of the spreading of the virus (Gaur, 2020). Nonetheless, although the situation was seemingly coming under control, a recent congregation at Delhi resulted in a huge spike in number of positive patients in the city, which was undoubtedly a dwindling phase for the Agra model.

It could prove to be a fatal stabbing to the containment of a virus at Agra in no matter of time. However, the diligent efforts of numerous healthcare professionals with an efficiency of 8.29% made the changes visible in the form of recovered patients. Unfortunately, they did face some uprising in mortality rate as it went over 2.48%. Still, despite feeling down, they doubled their dedication and unturned every stone leading to the situation getting normal (for COVID-19 statistics, please refer Appendix – Table 1.B, S. No. 5).

1.4.2.2 Worst-Case Scenarios

- **Jaipur Model**
  - Jaipur, the pink city, has been named as one of the worst affected areas due to COVID-19 and has registered itself as a hotspot in the country. India, which encountered traces of COVID-19 in a person with a history of foreign travel, instantly advised all the people returned from abroad to be in self-isolation for 14 days. Unfortunately, a 45-year-old Oman-returned individual stepped out of his home before the prescribed time. Over 94 individuals were detected positive, which connected with the patient zero, who was the first diagnosed positive case of COVID-19 in the city (Bohra, 2020). The situation could have been saved from going down this damaging path if the administration and policymakers had ensured zero mobility of the particular person returning from an earlier visit. Still, until then, the damage was already done. In addition to that, the targeted audience was less accurate and 2531.37 tests per million resulted in only 5.31% being positive. The population at home who delayed acceptance in understanding the fragility of the scenario made the danger worse and eventually made the administration take stricter and stiff actions against the violators.

With the passage of time, the administration and policymakers underwent a precise analysis and subsequently served people much better than before.

The severe condition of Jaipur could either burst out in destroying all the efforts of the three quadrants or be reduced by the accumulation of all the efforts of those three quadrants by the most respected profession, healthcare. But the doctors and their staff emerged in firefighting mode to combat COVID-19. The world admired the bravery of healthcare professionals at SMS Hospital in curing outpatients with a combination of medication, but the increasing number of patients dragged down their efficiency to 5.70%. As a result, 91.8% of active cases resulted from a heavy spike in affected people (TOI, 2020a). Meanwhile, the 2.47% mortality rate was observed mainly due to patients' previous poor medical history, which made healthcare professionals work more briskly to save the maximum number of patients from the clutches of COVID-19 (for COVID-19 statistics, please refer to Appendix – Table 1.B, S. No. 3).

- **Indore Model**
  - A city always bustling with commercial and socio-cultural activities, Indore has turned into the coronavirus infection hotspot in Madhya Pradesh. Ironically, it was ranked the cleanest city in the country for the third time in a row last year in the central government's cleanliness survey. According to the data released by the Chief Medical Health Officer's (CMHO) office, the number of cases in Indore accounts for more than half of the total cases in the state (PTI, 2020a).

    The administration and policymakers were doing their maximum to contain the spread of the disease. However, the situation was seemingly impossible to curb, and henceforth the corona case positivity rose to a high 22.2% in the city. The administration conducted 974.12 tests per million in no time, but the scenario didn't improve much. When the rest of India was planning on fighting the pandemic, Madhya Pradesh was going through a political turmoil that eventually led to a delayed response regarding the seriousness of COVID-19 (The Print Team, 2020).

    Nonetheless, the administration of strategies and methods for protecting the city can't be fruitful without the support of its citizens. A systematic and healthy response of these people to the situation and their well-being would have turned the situation upside down and saved so many lives (Dwary & Nair, 2020).

    All the healthcare professionals worked throughout the day to cure the patients, but unfortunately, the enormity of the situation made the mortality rate grow to 5.43%. However, the doctors didn't lose hope, and in counter-attack, they recovered patients with an efficiency of 7.99% and are still fighting on edge against COVID-19. They could locate the hotspots that made their testing prove to be 86.57% successful in determining active cases out of the population tested (for COVID-19 statistics, please refer Appendix – Table 1.B, S. No. 1).

- **Mumbai Model**
  - Mumbai, one of the most populated cities and regarded as the commercial capital of the country, engulfed with the movement of people throughout the day, came to a standstill under the wrath of the COVID-19 outbreak. Moreover, the situation worsened when Mumbai residents returned from various Middle East countries, sadly not alone but with traces of the deadly virus (TNN, 2020b). The administration and policymakers

waited for orders from the central government to seal the borders for restricting the inter-district movements. The same was done for testing, which was delayed and caused the infection to spread out to the masses, leading to 9.36% corona case positivity in the area.

With the help of flying squads, door-to-door surveys were carried out in all 227 municipal wards of Mumbai to identify suspected patients (TOI, 2020b). But, according to the complaint reported by patients, it was found that health workers could not carry out proper testing, which was unquestionably a matter of concern for curbing the spread of COVID-19. Amid difficulty in getting essential services, e-commerce delivery persons (third quadrant of COVID-19 warriors) selflessly delivered required commodities, food, pharmaceuticals, and medical equipment (ET Bureau, 2020b). NGOs and the residents, too, shared a helping hand to the administration in providing food to the needy.

Besides, rumor-mongers were found guilty of spreading false messages, which unfortunately accounted for gathering many migrant laborers near Bandra station in a day (Joshi, 2020). As a result, there was a high rise in the cases, making the administration take strict action to control the scenario. The administration might prevent such chaotic situations in the future if they release a communication channel between the government and ordinary people so that the spread of such false news could be controlled to some extent.

With the mortality rate gone above 4.98%, Mumbai's healthcare professionals, with all their might, fought against the virus. They achieved an efficiency of 11.19%, which is praiseworthy considering the significant number of positive cases. Leaving no stone unturned, they conducted 1312.99 tests per million to flatten the curve and were successful to a great extent (for COVID-19 statistics, please refer Appendix – Table 1.B, S. No. 2).

1.4.2.3 Comparative Discussion for Districts

Numbers unbiasedly show the accurate situation, which might be in polar opposition to the existing or popularized opinion. In the research, some unexpected and unusual results were obtained, which are listed later. After plotting the rate of recovery against corona case positivity, Indore can be classified as the worst affected city with over 22% positive cases to only 7.9% recovered. According to a report (Dwary & Nair, 2020), the maximum number of cases was discovered in Khajrana, Chandannagar, Ranipura, Tatpatti Bakhal, and Silawatpura, which had numerous people returning from the Delhi congregation who unknowingly spread the infection. Numbers designated against Agra signifies that the acclaimed central model has a mere recovery rate of 8.29%, which is even lower than one of the worst affected, Mumbai, with 11.19%. Bhilwara can be seen coming out of the clutches of COVID-19 with the highest recovered rate of 86.20%, mainly due to a ruthless lockdown and tremendous efforts by all COVID-19 warriors (Iqbal, 2020).

Something unexpected was achieved while plotting the rate of recovery against the mortality rate in Bhilwara, with the highest mortality rate than worst affected

Indore, Mumbai, and the rest, which the administration stated to be because of the poor medical history of patients (Parihar, 2020). Kasaragod doctors deserve a bow to not let any patient lose his or her life and to achieve a 0% mortality rate, which is the lowest. Undoubtedly, Jaipur has a high COVID-19 case positivity of 5.31%, but the four quadrants' countermeasures made it achieve the highest tests per million of 2531.37 among all (PTI, 2020c). The administration is leaving no stone unturned to find out the infected by rigorous testing in the city.

Indore had a two-sided attack with the lowest tests per million of 974.14 and with the highest corona case positivity of 22.20%. Therefore, it is an alarming situation for all quadrants of society to double their efforts to achieve victory over the deadly virus. As the graph plotted between rate of recovery to active cases shows, it can easily be seen that Jaipur had the highest number of active cases with 91.82% to only 5.70% recovered ones. Unfortunately, Agra had a higher percentage of active patients, with 89.21, than did the severely affected Mumbai and Indore. In no time, the situation could result in a disaster if more impactful measures aren't taken up by the Agra administration (Qureshi, 2020a). Reports stated that the sudden spike in the number of active cases was mainly because of residents of Agra returning from the congregation at Delhi.

1.4.2.4 Issues to Be Addressed for Districts

Outcomes of research for Agra and Indore brought some unwanted predictions of community spread in the cities, which is undoubtedly an alarming situation. Agra, which was able to curb the virus, is also slowly being caught in the shackles of COVID-19. This raises a question for authorities – how did they let the situation get away from them? Also needed are some possible suggestions, such as staying indoors. On the other hand, Indore is going through the worst phase, but fortunately, the study showed that the root of the problem was that the area had the lowest testing per million among all cities.

Considering Bhilwara's and Kasaragod's cluster mapping techniques that discerned between the genuinely positive and the suspicious, one can be acclaimed as a boon for Indore too. In no time, Indore, after doing so, will stand tall against COVID-19 and unquestionably become a role model for the whole nation. Jaipur can be seen as fighting the situation with full zeal and enthusiasm by testing out the maximum number of infected. Still, the real herculean task is bringing back those positive patients from the effects of COVID-19, as Jaipur does have a significant mortality rate which in no time could be reversed. Suggestions for more healthcare staff to prevent people from losing their lives could be a game-changer, since many reports stated the lack of ventilators and other items at Jaipur causing difficulty for healthcare professionals to help patients recover (TNN, 2020a).

The havoc in Mumbai is increasing day by day. The nightmare of community spread could be true anytime as the spread of the virus in the densely populated Dharavi area is reported. The destruction predicted can be reduced greatly if the number of mobile outpatient departments (OPDs) and testing kits increases more rapidly than the corona case positivity. Unfortunately, gathering a large number of people at stations questions the seriousness of the situation by the administration. Although the administration is undoubtedly trying its best to prevent it from being repeated, suggestions for implementing a strict screening for all the fake news and rumors and urging people to look out for official statements by the government could help to a great extent.

1.5 CONCLUSIONS

After all the research, it has been found that some cities are affected worse than states are. Statistics bluntly revealed that Mumbai, Indore and Bhopal, and Ahmedabad are infected with above 80% COVID-19 cases in the states respectively, while Jaipur, Agra, Bhilwara, and so on are among the many districts infected.

This research presents a framework for the administration and policymakers by highlighting the instances at various locations of the country that are steadily coming out of the crisis or are being more heavily affected. The situation in India for COVID-19 is very dynamic and challenging. Reports stated the discovery of 100 new cases in a day to no new cases found. That's where the limitation of this study was felt. Since, the current study had been carried out till mid-April 2020 and considering the dynamics in increase/decrease of COVID-19 cases, it would be good, if the COVID-19 trend is analyzed based on the utilized framework in a timeline manner, as it will help the concerned authorities take suitable measures to monitor the situation and take necessary measures to minimize COVID-19.

Optimistic approaches towards this study can convert an out of control situation into a controlled one since it contains challenges, shortcomings, suggestions, and questions that can be used for further research bringing out some evolutionary measures for saving humanity. On the positive side, it is a challenge and opportunity for researchers to develop technological innovations to help organizations dealing with crises currently and after the conclusion of this pandemic – further, numerous policy interventions along with effective collaborations among organizations, NGOs, and government authorities are currently needed to get the organizations back on the right track and curb the impact of COVID-19.

Financial support and Sponsorship: Not Applicable
Conflicts of interest: Nil

REFERENCES

Bhatia, S., & Devulapalli, S. (2020, April 19). Mapped: The spread of coronavirus across India's districts. *Live Mint.* https://www.livemint.com/news/india/mapped-the-spread-of-coronavirus-across-india-s-districts-11587179250870.html

Bhatt, R. (2020, April 26). The Bhilwara model. *THE WEEK.* https://www.theweek.in/theweek/statescan/2020/04/17/the-bhilwara-model.html

Bohra, S. (2020, April 9). 'Super spreader' infects nearly 100 people in Jaipur, Rajasthan gets second Covid-19 hotspot. *The Print.* https://theprint.in/india/super-spreader-infects-nearly-100-people-in-jaipur-rajasthan-gets-second-covid-19-hotspot/

Bruin, Y. B. de, Lequarre, A.-S., McCourt, J., Clevestig, P., Pigazzani, F., Jeddi, M. Z., Colosio, C., & Goulart, M. (2020). Initial impacts of global risk mitigation measures taken during the combatting of the COVID-19 pandemic. *Safety Science*, *128*, 1–8. https://doi.org/10.1016/j.ssci.2020.104773

Chaturvedi, H. (2020, April 12). Survey, identify, quarantine: A look at the Agra model in Covid-19 fight. *Hindustan Times*. https://www.hindustantimes.com/india-news/survey-identify-quarantine-a-look-at-the-agra-model-in-covid-19-fight/story-mQemI8jQatmg-DicZf7WNZK.html

Dwary, A., & Nair, A. (2020, April 11). Indore new coronavirus hotspot in Madhya Pradesh, records 72% of state deaths. *NDTV*. https://www.ndtv.com/cities/covid-19-indore-new-coronavirus-hotspot-in-madhya-pradesh-records-72-of-state-deaths-2210002

ET Bureau. (2020a, April 16). Mumbai to prioritise testing high-risk contacts with symptoms, move raises questions. *The Economic Times*. https://economictimes.indiatimes.com/news/politics-and-nation/mumbai-to-first-test-high-risk-contacts-with-symptoms/articleshow/75169149.cms?from=mdr

ET Bureau. (2020b, April 23). Coronavirus lockdown: Maharashtra revokes relaxations for Mumbai and Pune. *The Economic Times*. https://economictimes.indiatimes.com/news/politics-and-nation/lockdown-maharashtra-revokes-relaxations-from-mumbai-and-pune/articleshow/75276735.cms?utm_source=contentofinterest&utm_medium=text&utm_campaign=cppst

FE Online. (2020, April 13). COVID-19: Three strategies for 3 hotspots! How Agra, Bhilwara, Pathanamthitta models work; details. *The Indian Express*. https://www.financialexpress.com/lifestyle/covid-19-three-strategies-for-3-hotspots-how-agra-bhilwara-pathanamthitta-models-work-details/1926630/

Gaur, V. (2020, April 14). Agra sees 35 new cases as admin says patients only in hotspots. *The Economic Times*. https://economictimes.indiatimes.com/news/politics-and-nation/agra-sees-35-new-cases-as-admin-says-patients-only-in-hotspots/articleshow/75130609.cms

Ghosh, A. (2020a, April 21). Explained: How Kerala's Kasaragod has fought coronavirus. *The Indian Express*. https://indianexpress.com/article/explained/kerala-coronavirus-cases-kasaragod-model-6371484/

Ghosh, H. (2020b, April 11). COVID-19: Data shows West Bengal's testing is the lowest among larger states. *The Wire*. https://thewire.in/government/west-bengal-covid-19-testing

Goh, Y., Chua, W., Lee, J. K. T., Ang, B. W. L., Liang, C. R., Tan, C. A., Choong, D. A. W., Hoon, H. X., Ong, M. K. L., & Quek, S. T. (2020). Operational strategies to prevent coronavirus disease 2019 (Covid-19) spread in radiology: Experience from a singapore radiology department after severe acute respiratory syndrome. *Journal of the American College of Radiology*, 1–7. https://doi.org/10.1016/j.jacr.2020.03.027

Gupta, R., & Madgavkar, A. (2020). *Getting ahead of coronavirus: Saving lives and livelihoods in India*. Mc Kinsey & Company. https://www.mckinsey.com/featured-insights/india/getting-ahead-of-coronavirus-saving-lives-and-livelihoods-in-india

HT Correspondent. (2020, April 18). Centre picks Kerala's Kasaragod's Covid-19 containment model for high praise. *Hindustan Times*. https://m.hindustantimes.com/india-news/centre-picks-kerala-s-kasaragod-covid-19-containment-model-for-high-praise-says-we-need-to-fight together/story-4PdpnoQT7QZ83h8CLe2UzK_amp.html

Iqbal, M. (2020, April 10). In Rajasthan's Bhilwara, 'ruthless containment' model breaks virus transmission chain. *The Hindu*. https://www.thehindu.com/news/national/other-states/in-rajasthans-bhilwara-ruthless-containment-model-breaks-virus-transmission-chain/article31307764.ece

Jayarajan, S. (2020a, March 25). In Kasaragod, virus spread from patient to 4 contacts within 20 minutes: Collector. *The News Minute*. https://www.thenewsminute.com/article/kasaragod-virus-spread-patient-4-%0D%0Acontacts-within-20-minutes-collector-121095%0A%0A%0A%0A

Jayarajan, S. (2020b, April 6). Student with no symptoms of COVID-19 tests positive after 19 days in Pathanamthitta. *The News Minute*. https://www.thenewsminute.com/article/student-no-symptoms-covid-19-tests-positive-after-19-days-pathanamthitta-121983

Joshi, S. (2020, April 14). Cops lathicharge migrants as thousands gather at Bandra station to leave Mumbai, defy lockdown orders. *India Today*. https://www.indiatoday.in/india/story/lockdown-woes-mumbai-stations-flooded-with-migrant-labourers-hoping-to-get-back-home-1666908-2020-04-14?fbclid=IwAR2YFzC7UE97cUckcOIIUJ-aPLheQT6DVxW19K7–6-TFjGgxZ6t0zmu3188

Koshy, S. M. (2020, April 13). As Kerala slows Coronavirus cases, how a hotspot turned its story around. *NDTV*. https://www.ndtv.com/kerala-news/covid-19-india-as-kerala-slows-coronavirus-cases-how-a-hotspot-turned-its-story-around-2211096

Mekaad, S., & Singh, A. (2020, April 15). How MP's two 'Swachh cities' Indore, Bhopal turned hotspots. *The Times of India*. https://timesofindia.indiatimes.com/city/bhopal/how-mps-two-swachh-cities-indore-bhopal-turned-hotspots/articleshow/75151624.cms

Mishra, A. R. (2020, April 14). Barclays cuts GDP forecast for India to zero for 2020. *Live Mint*. https://www.livemint.com/news/india/barclays-cuts-gdp-forecast-for-india-to-zero-for-2020-11586848023858.html

Parihar, R. (2020, March 26). Rajasthan debates cause of death as Covid-19 patient dies in Bhilwara. *India Today*. https://www.indiatoday.in/coronavirus-outbreak/story/rajasthan-debates-cause-of-death-as-second-covid-19-patient-dies-in-bhilwara-1660106-2020-03-26

PTI. (2020a, April 9). Indore's journey from cleanest city to coronavirus hotspot. *India Today*. https://www.indiatoday.in/india/story/indore-journey-from-cleanest-city-to-coronavirus-hotspot-1665026-2020-04–09

PTI. (2020b, April 10). Gujarat reports 241 coronavirus cases, 17 deaths. *Economic Times*. https://m.economictimes.com/news/politics-and-nation/55-new-coronavirus-cases-reported-in-gujarat-state-tally-241/amp_articleshow/75060580.cms

PTI. (2020c, April 11). Rajasthan records 139 new COVID-19 cases; 25,000 people to be tested in single day. *The Economic Times*. https://economictimes.indiatimes.com/news/politics-and-nation/18-more-found-positive-for-covid-19-in-rajasthan-number-rises-to-579/articleshow/75091335.cms?from=mdr

PTI. (2020d, April 17). Coronavirus infects Indore's image amid rising cases, deaths. *India Today*. https://www.indiatoday.in/india/story/coronavirus-infects-indore-s-image-amid-rising-cases-deaths-1668157-2020–04–17

Qayam. (2020, April 19). 45 new corona cases in Taj city, tally reaches 241. *The Siasat Daily*. https://www.siasat.com/45-new-corona-cases-taj-city-tally-reaches-241-1876360/

Qureshi, S. (2020a, April 12). Coronavirus in India: Over 50% of Covid-19 cases in Agra related to Tablighi Jamaat. *India Today*. https://www.indiatoday.in/india/story/coronavirus-in-india-over-covid-19-cases-agra-related-to-tablighi-jamaat-1666155-2020-04-12

Qureshi, S. (2020b, April 23). Agra administration has failed people in containing Covid-19: Local leaders. *India Today*. https://www.indiatoday.in/india/story/agra-administration-has-failed-people-in-containing-covid-19-local-leaders-1670294-2020-04-23

Roy, D., Tripathy, S., Kar, S. K., Sharma, N., Verma, S. K., & Kaushal, V. (2020). Study of knowledge, attitude, anxiety & perceived mental healthcare need in Indian population during COVID-19 pandemic. *Asian Journal of Psychiatry*, *51*, 1–7. https://doi.org/10.1016/j.ajp.2020.102083

Roy, D. D. (2020, April 11). Coronavirus – 30 COVID-19 hotspots now in Delhi, cases rise to 903. *NDTV*. https://www.ndtv.com/delhi-news/coronavirus-covid-19-delhi-30-hotspots-now-in-delhi-cases-rise-to-903–2209837

SCC. (2020, April 9). *Agra launches mobile app to keep a tab on Covid-19 lockdown*. Smart Cities Council. https://india.smartcitiescouncil.com/article/agra-launches-mobile-app-keep-tab-covid-19-lockdown

Sharma, T. (2020, April 11). The heroes of Bhilwara and the story of their war against COVID-19. *Money Control*. https://www.moneycontrol.com/news/trends/health-trends/the-bhilwara-model-those-behind-the-successful-containment-of-covid-19–5134601.html

The Hindu Data Team. (2020, April 3). COVID-19 | State-wise tracker for coronavirus cases, deaths and testing rates. *The Hindu*. https://www.thehindu.com/data/covid-19-state-wise-tracker-for-coronavirus-cases-deaths-and-testing-rates/article31248444.ece

The Print Team. (2020, April 13). Has politics in Madhya Pradesh damaged its battle against coronavirus? *The Print*. https://theprint.in/talk-point/has-politics-in-madhya-pradesh-damaged-its-battle-against-coronavirus/400788/?amp

TNN. (2020a, March 24). Hosps have only 1,500 ventilators. *The Times of India*. https://timesofindia.indiatimes.com/city/jaipur/hosps-have-only-1500-ventilators/articleshow/74783592.cms

TNN. (2020b, April 18). 93% drop in a day? Only 12 cases in Mumbai, says government. *The Times of India*. https://timesofindia.indiatimes.com/city/mumbai/93-drop-in-a-day-only-12-cases-in-city-says-govt/articleshow/75212188.cms

TOI. (2020a, March 21). Rajasthan doctors cure coronavirus patient with HIV drugs. *Times of India*. https://m.timesofindia.com/city/jaipur/city-docs-cure-corona-patient-with-hiv-drugs/amp_articleshow/74584859.cms

TOI. (2020b, April 1). Covid-19: Maha govt sets up flying squads to conduct door-to-door survey in Mumbai. *The Times of India*. https://timesofindia.indiatimes.com/videos/city/mumbai/covid-19-maha-govt-sets-up-flying-squads-to-conduct-door-to-door-survey-in-mumbai/videoshow/74925203.cms

Waris, A., Atta, U. K., Ali, M., Asmat, A., & Baset, A. (2020). COVID-19 outbreak: Current scenario of Pakistan. *New Microbes and New Infections*, *35*, 1–6. https://doi.org/10.1016/j.nmni.2020.100681

WHO. (2020). *Q&A on coronaviruses (COVID-19)*. World Health Organization. https://www.who.int/news-room/q-a-detail/q-a-coronaviruses

World Population Review. (2020). *2020 world population by country*. https://worldpopulationreview.com/

APPENDIX

TABLE 1.A
COVID-19 Statistics for Selected States

S. No	State	Tested Population	Cases Confirmed	Cases Recovered	Deaths	Corona Case Positivity (%)	Rate of Recovery (%)	Mortality Rate (%)	Test per Million People	% Active Cases
1	Delhi	18,606	1707	72	42	9.1745	4.2179	2.4605	1060	93.3216
2	Madhya Pradesh	15,302	1310	69	75	8.5610	5.2672	5.7252	182.5	89.0076
3	West Bengal	4212	287	55	10	6.8139	19.1638	3.4843	42.7	77.3519
4	Telangana	10,992	766	186	18	6.9687	24.2820	2.3499	282.4	73.3681
5	Maharashtra	60,284	3323	331	201	5.5122	9.9609	6.0488	494.4	83.9904
6	Jammu and Kashmir	6438	328	42	5	5.0947	12 8049	1.5244	478	85.6707
7	Gujarat	23,438	1272	88	48	5.4271	6.9182	3.7736	361.7	89.3082
8	Tamil Nadu	29,673	1323	283	15	4.4586	21.3908	1.1338	384.5	77.4754
9	Punjab	5324	211	27	14	3.9632	12.7962	6.6351	178.2	80.5687
10	Uttar Pradesh	21,179	849	86	14	4.0087	10.1296	1.6490	90.7	88.2214

TABLE 1.B
COVID-19 Statistics for Selected Districts

S. No	District	Tested Population	Cases Confirmed	Cases Recovered	Deaths	Corona Case Positivity (%)	Rate of Recovery (%)	Mortality Rate (%)	% Active Cases	Total Population	Tests per Million People
1	Indore	4057	901	72	49	22.2085	7.9911	5.4384	86.5705	4,164,692	974.1417
2	Mumbai	26,800	2509	281	125	9.3619	11.1997	4.9821	83.8183	20,411,274	1312.9999
3	Jaipur	9896	526	30	13	5.3153	5.7034	2.4715	91.8251	3,909,333	2531.3781
4	Bhilwara	5721	29	25	2	0.5069	86.2069	6.8966	6.8966	452,209	12,651.2299
5	Agra	3891	241	20	6	6.1938	8.2988	2.4896	89.2116	2,210,246	1760.4375
6	Kasaragod	2600	167	117	0	6.4231	70.0599	0	29.9401	1,381,399	1882.1499

2 Production-Inventory Model for Perishable Items under COVID-19 Pandemic Disruptions

Vikash Murmu, Dinesh Kumar, and Biswajit Sarkar

CONTENTS

2.1 INTRODUCTION AND BACKGROUND

The severe acute respiratory syndrome coronavirus 2 (SARS-CoV-2) that affects humans' upper and lower respiratory tract is named the COVID-19 pandemic that affected more than 200 countries worldwide, causing massive human casualties and economic losses. In addition, the pandemic has pushed the world towards a global recession scenario by creating a series of disruptions in the supply chain network. To contain this pandemic, the government has to impose lockdown and quarantine orders to stop the further spread of the virus. Consequently, the pandemic resulted in the shutdown of various industries, causing a significant supply chain disruption that has seen international trade decrease between 13% and 32% in 2020–21(WTO, 2020).

The whole world has come to a standstill after the arrival of SARS-COV-2. This virus is highly contagious (Christidis & Christodoulou, 2020); therefore, governments had to impose a lockdown in the state to contain the virus, which disrupts the whole supply chain network (Basu, 2020a). These disruptions in the supply chain have created certain challenges for organizations as follows:

- The demand for essential products such as foods and medicines has increased, and the need for non-essential products has decreased (Paul &

DOI: 10.1201/9781003150084-2

Chowdhury, 2021). This increase in demand for certain items leads to partial shortages of these products in the market (Deaton & Deaton, 2020). This demand spike is due to the panic purchasing behaviour of people, insecurity about the future, and stockpiling of products (Hobbs, 2020). In addition, this change in people's purchasing behaviour is due to the observed threats about post-pandemic situations and copying of people's stockpiling attitude that occurred just after the lockdown was relaxed (Yuen et al., 2020).

- The demand for non-essential items has fallen as the income of people goes down. Therefore, to hold their savings for the future, they prefer to spend it on essential goods rather than non-essential ones (Chiaramonti & Maniatis, 2020).
- People's income has decreased due to the closure of tourism industries (Majumdar et al., 2020), which is one of the significant contributors to GDP for any country. The closure of the tourism industry highly impacted the income of hotels and restaurants, resulting in job losses.
- The unforeseeable demand disruption has created many challenges in forecasting and decision-making (Gunessee & Subramanian, 2020).
- This pandemic has caused severe production disruptions due to the uncertainty in workforce availability (Leite et al., 2021), raw material, and logistics, causing an enormous backlog (Richards & Rickard, 2020).
- The international supply chain network, transportation, and logistics management was disrupted due to the closure of air, water, rail, and road transportation systems, causing various losses for organizations dealing in international trade (Govindan et al., 2020).
- Social distancing and isolation caused by pandemics have resulted in ambiguity in the supply chain as it disrupts the interaction among the various supply chain partners resulting in a loss of collaborative efforts (Baveja et al., 2020), (Gunessee & Subramanian, 2020), (Remko, 2020).
- The lockdown has disrupted the food supply chain network and resulted in food insecurity (Id & Khatun, 2021). Due to the lockdown situation, the prices of various crops such as wheat and maize have fallen during the early 2020s ("Covid-19 Pandemic – Impact on Food and Agriculture," 2019), and because of this, the income of food-service sectors such as restaurants and hotels declines (Brinca et al., 2020). The closure of these food-service sectors resulted in a loss in the supply chain, including farmers, the primary producers. The perishable food industry faces significant challenges in keeping its products fresh during the lockdown period. Furthermore, the travel restriction has prevented farmers from accessing the open market, pushing them towards reducing crop productivity yield ("Covid-19 Pandemic – Impact on Food and Agriculture," 2019). It has also put immense pressure on the warehousing activities due to surplus inventories, especially perishable ones (Bochtis et al., 2020).
- The travel restriction has also impacted the agricultural workforce as they were stopped from migrating to other states for farming purposes during the farm season resulting in lower productivity of crops (Benos et al., 2020a, 2020b). However, a labor shortage during the harvesting season has

> disrupted the production yield, making the market more volatile (Fortune & Foote, 2020). Moreover, this disruption in crop production also lowered the use of pesticides and fertilizers, creating a loss for these industries (Jámbor et al., 2020). Thus, the entire production chain from fertilizer to crop yield has been affected by the COVID-19 pandemic.

These disruptions within the supply chain have crashed the whole network, including the warehouses. As the lockdown period prolongs, the warehoused perishable products such as fruits, vegetables, dairy products, and meat lose their value due to faster deterioration and become unusable during this period. It creates waste in terms of revenue for the organization (Mor et al., 2020). Therefore, a robust inventory model should be developed to assist inventory managers in optimizing the ordering quantity, cycle time, and selling price to maximize the firm's profit. In this regard, researchers have tried to make a prototype model that takes the COVID-19 pandemic into consideration, with the demand disruption and uncertainty in the lockdown period (Rana et al., 2021; Ivanov, 2020).

The COVID-19 pandemic has created a state of fear in people's minds (Barkur & Kamath, 2020), as several people have lost their jobs during the pandemic, and now they fear losing the availability of foods stocks for their families. Therefore, when the government lifted the lockdown with certain restrictions, people desperately started purchasing food items to store them for post-pandemic situations (Bochtis et al., 2020). It resulted in a sudden increase in demand for these items following the emptying of supermarket shelves. The pandemic has also changed customers' food habits, as now they are opting more for stable products such as rice and wheat rather than non-stable ones such as fast foods (Vancic & Pärson, 2020). This panic purchasing behaviour is for the short term because, in the longer run, the demand for certain perishable products will fall gradually. Therefore, in the future, the demand for perishable items may become more price sensitive, with people purchasing fewer expensive food items.

In the past, certain resilient techniques have been invented to cope with disruptions; however, the disruption caused by this pandemic has overshadowed others. The COVID-19 pandemic has shown to the world the worst scenario, in which the availability of critical stock collapses (Amjath Babu et al., 2020) with the sudden surge in demand caused by the change in customers' purchasing behaviour (Brinca et al., 2020) and shortages in raw material due to disruption in the supply chain (Toffolutti et al., 2020). In previous papers, disruption has been defined as the amalgamation of the accidental and unpredicted event that affects the supply chain network (Sarkar & Kumar, 2015). The disruptions can either be natural or artificial events, such as labor union strikes, fire outbreaks, and wars between countries or states. In addition, some disruptions include floods, earthquakes, and extreme weather conditions (Jabbarzadeh et al., 2016). Still, the disruption caused by the COVID-19 pandemic is of unprecedented magnitude. At present, the closest known resilient techniques for disease outbreaks focus mainly on the medical supplies. The disruptions generally considered in the previous papers are (i) economic recession, (ii) market fluctuation, (iii) physiological panic among customers, and (iv) natural calamities (Bhattacharya et al., 2013).

On the other hand, very few studies have focused on the perishable foods chain disruption during a large disease outbreak (Golan et al., 2020). To avoid such scenarios in the future, it is required to expand the scope of the research work towards unknown disruptions with different factors and considerations. Therefore, a model should be prepared to replicate these scenarios so that a study can be conducted and the future can benefit (Rana et al., 2021). The recent literature with gaps is presented in Table 2.1.

The past literature mainly focuses on the theoretical aspect of COVID-19 disruption, but this is insufficient to measure the exact damage caused by this pandemic; hence, a mathematical model should be formulated to study this disruption concerning cost and revenue profit involved. Therefore, this chapter proposes a mathematical model considering the COVID-19 pandemic, such as disruptions in production and demand for the perishable inventory kept in the warehouse. This study aims to minimize the total cost and provide suggestive measures to make a resilient supply chain network.

The organization of this chapter is as follows: Section 2.2 states the assumptions and notations; model description has been explained in Section 2.3; the numerical example is shown in Section 2.4; Section 2.5 investigates the model's behaviour through sensitivity analysis; and results and discussions are presented in Section 2.6 followed by the conclusion and future work.

TABLE 2.1
Literature Reviews and Their Classifications

Papers	Production disruption	Demand disruption	COVID-19 pandemic	Resilience techniques	Warehouse management	Perishable products
(Christidis & Christodoulou, 2020)			✓			
(Basu, 2020b)			✓			
(Paul & Chowdhury, 2020)	✓	✓	✓	✓		
(Chiaramonti & Maniatis, 2020)			✓	✓		
(Majumdar et al., 2020)	✓		✓	✓		
(Richards & Rickard, 2020)		✓	✓	✓		✓
(Rana et al., 2021)		✓	✓	✓	✓	✓
(Govindan et al., 2020)		✓	✓	✓		✓
(Baveja et al., 2020)			✓	✓		
(Mor et al., 2020)	✓		✓	✓		✓
(Id & Khatun, 2021)	✓		✓	✓		✓
(Jámbor et al., 2020)	✓	✓	✓	✓		✓
This chapter	✓	✓	✓	✓	✓	✓

2.2 ASSUMPTIONS

1. The demand rate is price sensitive, i.e., $D_O = kS^{-e}$, where k is the scale parameter, S is the selling price, and e is the price elasticity (Rana et al., 2020).
2. The planning horizon is considered infinite. The replenishment rate is continuous.
3. The deterioration rate is constant, and the storage capacity in the warehouse is finite.
4. The cycle time is variable, and the unit holding is constant. The production process is assumed to be flawless, i.e., the probability of manufacturing/processing defective goods is negligible.

NOTATIONS

Z(t) = The level of inventory at any time t
θ = Deterioration rate
P = Production rate
D_O = Demand rate
ΔP = Disrupted production rate
ΔP_1 = Increased production rate
Δd = Disrupted demand rate
Δd_1 = Increased demand rate
t_L = Time at which the lockdown starts
t_O = Time at which the lockdown ends
t_1 = Time at which the inventory becomes maximum
t_2 = Time at which the inventory gets exhausted
H = holding cost/unit/time
c = Purchasing cost/unit/time
L = Maximum inventory

2.3 MODEL DESCRIPTION

In this chapter, two different scenarios have been considered with different lockdown periods. In both scenarios, the lockdown period disrupts the production as well as the demand. In both scenarios, a single warehouse system has been considered to minimize its total cost.

2.3.1 SCENARIO 1

This scenario has been studied in the time interval 0 to t_2 with a constant deterioration rate. The inventory starts building up during the interval 0 to t_1, and is then depleted from t_1 to t_2. From 0 to t_L, the scenario begins with normal production and demand situation. The production and demand get disrupted by ΔP and Δd respectively due to enforcement of lockdown in the time interval t_L to t_O. After the lockdown period,

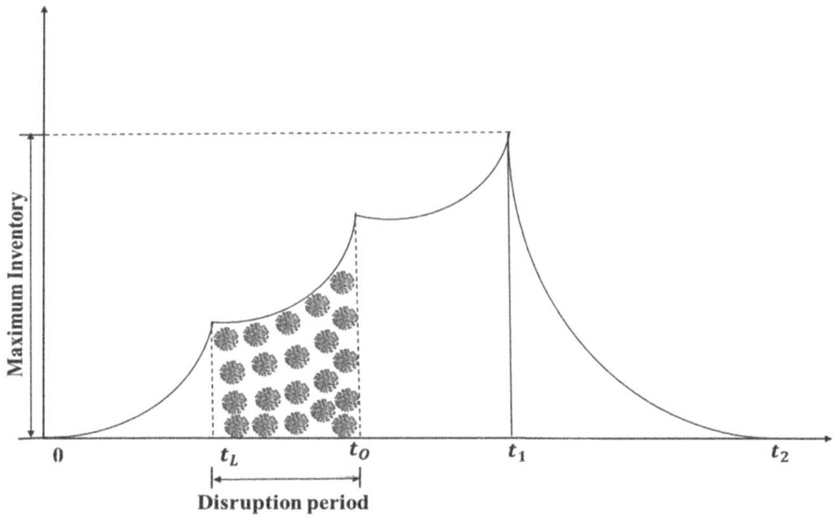

FIGURE 2.1 Production and dispatching of goods in lockdown scenario 1.

the government allows the movement of people with certain restrictions, then the production and demand increase by ΔP_1 and Δd_1 as shown in time interval t_0 to t_1. This is due to the panic purchasing behaviour of the people. At t_1, the production is stopped as it reaches the maximum inventory level. From t_1 to t_2 the inventory starts decreasing, and it becomes nil at t_2. This is shown graphically in Figure 2.1.

From 0 to t_L, inventory starts building up in a warehouse following the given differential equation:

$$\frac{dZ(t)}{dt}+\theta Z(t)=P-D_0 \quad (0 \le t \le t_L) \tag{2.1}$$

Taking the initial boundary condition t = 0 and Z(t) = 0, the equation becomes:

$$Z(t)=\left(\frac{P-D_0}{\theta}\right)\left(1-\frac{1}{e^{\theta t}}\right) \tag{2.2}$$

From t_L to t_O the production and demand are disrupted due to imposition of the lockdown by the government. The governing equation for this is:

$$\frac{dZ(t)}{dt}+\theta Z(t)=P-\Delta P-(D_0-\Delta d) \quad (t_L \le t \le t_0) \tag{2.3}$$

Taking the boundary condition at $t = t_L$, The equation becomes:

$$Z(t)=\frac{\left(P-\Delta P-(D_0-\Delta d)\right)}{\theta}+e^{\theta t_L}\left(\left(\frac{P-D_0}{\theta}\right)\left(\frac{-1}{e^{\theta t_L}}\right)+\frac{(\Delta P+\Delta d)}{\theta}e^{\theta t_L}\right)\frac{1}{e^{\theta t}} \tag{2.4}$$

During the time interval t_O to t_1, the governments lift the lockdown with certain restrictions, and due to this, the production and the demand increase by ΔP_1 and Δd_1. This increase in production and demand is caused by panic purchasing behaviour of the customers, as discussed earlier. Therefore, the governing equation in this time interval is:

$$\frac{dZ(t)}{dt}+\theta Z(t)=P+\Delta P_1-(D_0+\Delta d_1)\ (t_0 \le t \le t_1) \tag{2.5}$$

Considering the boundary condition at $t = t_O$, equation (2.5) yields:

$$Z(t)=\frac{(P+\Delta P_1-(D_0+\Delta d_1))}{\theta}+\left\{\left[\frac{\Delta P_1-\Delta d_1-\Delta P+\Delta d}{\theta}\right]e^{\theta t_0}\right.$$
$$\left.-\left[\left(\frac{P-D_0}{\theta}\right)\left(\frac{1}{e^{\theta t_L}}\right)+\left(\frac{\Delta P+\Delta d}{\theta}\right)e^{\theta t_L}\right]e^{\theta t_L}\right\}\frac{1}{e^{\theta t}} \tag{2.6}$$

At $t = t_1$ and $Z(t_1) = L$, we get:

$$t_1=\frac{1}{\theta}log\left\{\frac{\left[\frac{\Delta P_1-\Delta d_1-\Delta P+\Delta d}{\theta}\right]e^{\theta t_0}-\left[\left(\frac{P-D_0}{\theta}\right)\frac{1}{e^{\theta t_L}}+\left(\frac{\Delta P+\Delta d}{\theta}\right)e^{\theta t_L}\right]e^{\theta t_L}}{L-\left[\frac{P+\Delta P_1-(\Delta d+\Delta d_1)}{\theta}\right]}\right\} \tag{2.7}$$

At $t = t_1$, the inventory reaches its maximum level. At this point, the production is stopped, and demand continuous until $t = t_2$. The governing equation in this period is:

$$\frac{dZ(t)}{dt}+\theta Z(t)=-(D_0+\Delta d_1)\ (t_1 \le t \le t_2) \tag{2.8}$$

Stating the boundary condition at $t = t_1$, equation (2.8) becomes:

$$Z(t)=\frac{-(D_0+\Delta d_1)}{\theta}+\left(Le^{\theta t_1}+(D_0+\Delta d_1)\frac{e^{\theta t_1}}{\theta}\right)\frac{1}{e^{\theta t}} \tag{2.9}$$

At $t = t_2$ and $Z(t_2) = 0$, we get:

$$t_2=\frac{1}{\theta}log\left\{\frac{\theta\left[L+\left(\frac{d+\Delta d_1}{\theta}\right)\right]e^{\theta t_1}}{(D_0+\Delta d_1)}\right\} \tag{2.10}$$

At $t = t_2$, the inventory is wholly consumed in the warehouse.

Now the value of variable costs is computed in the interval 0 to t_2.

(i) Holding cost includes all the expenses that arose from holding the item in the inventory:

$$HC_{OW} = H\left\{\int_0^{t_L} Z(t)\,dt + \int_{t_L}^{t_0} Z(t)\,dt + \int_{t_0}^{t_1} Z(t)\,dt + \int_{t_1}^{t_2} Z(t)\,dt\right\} \tag{2.11}$$

Suppose: $HC_{OW} = H\{L_1 + L_2 + L_3 + L_4\}$

$$L_1 = \int_0^{t_L} Z(t)\,dt = H\left(\frac{P - D_0}{\theta}\right)\left(t_L + \frac{\left(e^{-\theta t_L} - 1\right)}{\theta}\right)$$

$$L_2 = \int_{t_L}^{t_0} Z(t)\,dt = H\left\{\left(\frac{P - \Delta P - (D_0 - \Delta d)}{\theta}\right)(t_0 - t_L) - e^{\theta t_L}\left[\left(\frac{P - D_0}{\theta}\right)\left(-\frac{1}{e^{-\theta t_L}}\right) + \left(\frac{\Delta P + \Delta d}{\theta}\right)e^{\theta t_L}\right]\left(\frac{e^{-\theta t_0}}{\theta} - \frac{e^{-\theta t_L}}{\theta}\right)\right\}$$

$$L_3 = \int_{t_0}^{t_1} Z(t)\,dt = H\left\{\left(\frac{P + \Delta P_1 - (D_0 + \Delta d_1)}{\theta}\right)(t_1 - t_0) - \left[\left(\frac{\Delta P_1 - \Delta d_1 - \Delta P + \Delta d}{\theta}\right)e^{\theta t_0} - \left(\left(\frac{P - D_0}{\theta}\right)\left(\frac{1}{e^{\theta t_L}}\right) + \left(\frac{\Delta P + \Delta d}{\theta}\right)e^{\theta t_L}\right)e^{t_L}\right]\left(\frac{e^{\theta t_L}}{\theta} - \frac{e^{\theta t_0}}{\theta}\right)\right\}$$

$$L_4 = \int_{t_1}^{t_2} Z(t)\,dt = H\left\{-\left(\frac{D_0 + \Delta d_1}{\theta}\right)(t_2 - t_1) - \left[Le^{\theta t_L} + (D_0 + \Delta d_1)\frac{e^{\theta t_1}}{\theta}\right]\left(\frac{e^{-\theta t_2}}{\theta} - \frac{e^{-\theta t_1}}{\theta}\right)\right\}$$

(ii) Deterioration cost, the cost experienced due to deterioration of items in the inventory, is given as follows:

$$DC = \theta c\left[\int_0^{t_L} Z(t)\,dt + \int_{t_L}^{t_0} Z(t)\,dt + \int_{t_0}^{t_1} Z(t)\,dt + \int_{t_1}^{t_2} Z(t)\,dt\right] \tag{2.12}$$

Suppose: $DC = \theta c\left[L_5 + L_6 + L_7 + L_8\right]$

$$L_5 = \theta c\int_0^{t_L} Z(t)\,dt = c\left(P - D_0\right)\left(t_L + \left(\frac{e^{-\theta t_L}}{\theta} - \frac{1}{\theta}\right)\right)$$

$$L_6 = \theta c\int_{t_L}^{t_o} Z(t)\,dt = \theta c\left[\left(\frac{P - \Delta P - \left(D_0 - \Delta d\right)}{\theta}\right)\left(t_0 - t_L\right)\right.$$
$$- e^{\theta t_L}\left(\left(\frac{P - D_0}{\theta}\right)\left(\frac{-1}{e^{\theta t_L}}\right) + \left(\frac{\Delta P + \Delta d}{\theta}\right)e^{\theta t_L}\right)$$
$$\left.\left(\frac{e^{-\theta t_o}}{\theta} - \frac{e^{-\theta t_L}}{\theta}\right)\right]$$

$$L_7 = \theta c\int_{t_o}^{t_1} Z(t)\,dt = \theta c\left\{\left(\frac{P + \Delta P_1 - \left(D_0 + \Delta d_1\right)}{\theta}\right)\left(t_1 - t_0\right)\right.$$
$$-\left[\left(\left(\frac{\Delta P_1 - \Delta d_1 - \Delta P + \Delta d}{\theta}\right)e^{\theta t_0}\right.\right.$$
$$\left.-\left(\left(\frac{P - D_0}{\theta}\right)\left(\frac{1}{e^{\theta t_L}}\right) + \left(\frac{\Delta P + \Delta d}{\theta}\right)e^{\theta t_L}\right)\right)e^{\theta t_L}\Bigg]$$
$$\left.\left(\frac{e^{-\theta t_1}}{\theta} - \frac{e^{-\theta t_o}}{\theta}\right)\right\}$$

$$L_8 = \int_{t_1}^{t_2} Z(t)\,dt = \theta c\left\{\left(-\frac{\left(D_0 + \Delta d_1\right)}{\theta}\right)\left(t_2 - t_1\right)\right.$$
$$\left.-\left[Le^{\theta t_1} + \left(D_o + \Delta d_1\right)\frac{e^{\theta t_1}}{\theta}\right]\left(\frac{e^{-\theta t_2}}{\theta} - \frac{e^{-\theta t_1}}{\theta}\right)\right\}$$

(iii) Purchasing cost, the cost for purchasing the goods to be stored in the inventory, is as follows:

$$PC = cL \tag{2.13}$$

All the relevant cost has been found out during the time interval 0 to t_2. Therefore, the total cost will be:

$$TC = HC_{rw} + DC + PC \tag{2.14}$$

Placing the values in equation (2.14), the total cost becomes:

$$TC(S,L)=c(P-D_0)\left(t_L+\left(\frac{e^{-\theta t_L}}{\theta}-\frac{1}{\theta}\right)\right)+$$

$$H\left\{\left(\frac{P-\Delta P-(D_0-\Delta d)}{\theta}\right)(t_o-t_L)-e^{\theta t_L}\left[\left(\frac{P-D_0}{\theta}\right)\left(-\frac{1}{e^{-\theta t_L}}\right)+\left(\frac{\Delta P+\Delta d}{\theta}\right)e^{\theta t_L}\right]\right.$$

$$\left.\left(\frac{e^{-\theta t_0}}{\theta}-\frac{e^{-\theta t_L}}{\theta}\right)\right\}+H\left\{\left(\frac{P+\Delta P_1-(D_0+\Delta d_1)}{\theta}\right)(t_1-t_0)-\left[\left(\frac{\Delta P_1-\Delta d_1-\Delta P+\Delta d}{\theta}\right)e^{\theta t_0}\right.\right.$$

$$\left.\left.-\left(\left(\frac{P-D_0}{\theta}\right)\left(\frac{1}{e^{\theta t_L}}\right)+\left(\frac{\Delta P+\Delta d}{\theta}\right)e^{\theta t_L}\right)e^{t_L}\right]\left(\frac{e^{\theta t_1}}{\theta}-\frac{e^{\theta t_0}}{\theta}\right)\right\}$$

$$+H\left\{-\left(\frac{D_0+\Delta d_1}{\theta}\right)(t_2-t_1)-\left[Le^{\theta t_1}+(D_0+\Delta d_1)\frac{e^{\theta t_1}}{\theta}\right]\left(\frac{e^{-\theta t_2}}{\theta}-\frac{e^{-\theta t_1}}{\theta}\right)\right\}$$

$$+c(P-D_0)\left(t_L+\left(\frac{e^{-\theta t_L}}{\theta}-\frac{1}{\theta}\right)\right)+\theta c\left[\left(\frac{P-\Delta P-(D_0-\Delta d)}{\theta}\right)(t_0-t_1)-e^{\theta t_L}\right.$$

$$\left.\left(\left(\frac{P-D_0}{\theta}\right)\left(\frac{-1}{e^{\theta t_L}}\right)+\left(\frac{\Delta P+\Delta d}{\theta}\right)e^{\theta t_L}\right)\left(\frac{e^{-\theta t_0}}{\theta}-\frac{e^{-\theta t_L}}{\theta}\right)\right]+\theta c$$

$$\left\{\left(\frac{P+\Delta P_1-(D_0+\Delta d_1)}{\theta}\right)(t_1-t_0)-\left[\left(\left(\frac{\Delta P_1-\Delta d_1-\Delta P+\Delta d}{\theta}\right)e^{\theta t_0}\right.\right.\right.$$

$$\left.\left.\left.-\left(\left(\frac{P-D_0}{\theta}\right)\left(\frac{1}{e^{\theta t_L}}\right)+\left(\frac{\Delta P+\Delta d}{\theta}\right)e^{\theta t_L}\right)\right)e^{\theta t_L}\right]\left(\frac{e^{-\theta t_1}}{\theta}-\frac{e^{-\theta t_0}}{\theta}\right)\right\}$$

$$+\theta c\left\{\left(-\frac{(D_0+\Delta d_1)}{\theta}\right)(t_2-t_1)-\left[Le^{\theta t_1}+(D_0+\Delta d_1)\frac{e^{\theta t_1}}{\theta}\right]\left(\frac{e^{-\theta t_2}}{\theta}-\frac{e^{-\theta t_1}}{\theta}\right)\right\}+cL$$

The objective is to minimize the total cost. Therefore, the required condition for this is:

$$\frac{d^2TC(S,L)}{dS^2}\geq 0 \text{ and } \frac{d^2TC(S,L)}{dL^2}\geq 0$$

For simplicity, the convexity has been proved graphically using the Matlab software, as shown in Figure 2.2.

ALGORITHM FOR SCENARIO 1:

Step 1: Initialize the input parameters.

Step 2: Initialize the number of items entering the inventory system [L] and selling price per unit of item [b].

Step 3: Input value of L and S. For example $[182 \leq L \leq 200]$ and $[15 \leq S \leq 30]$.

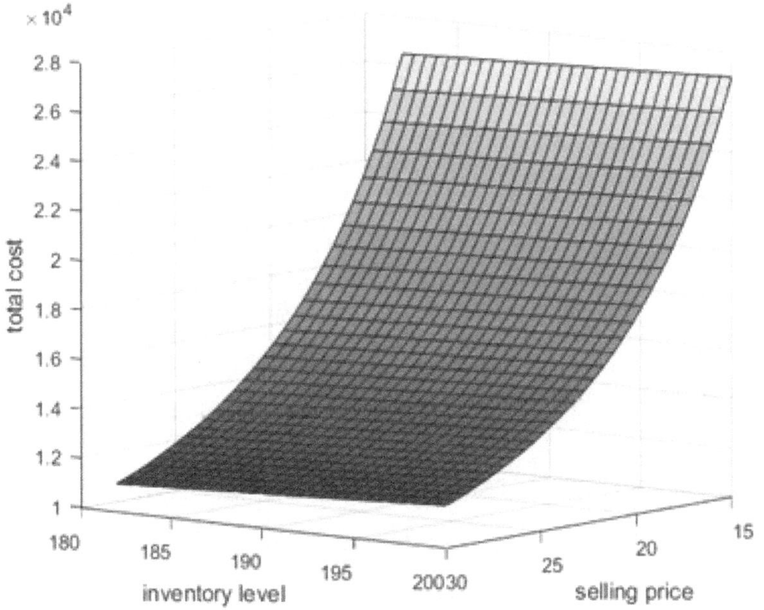

FIGURE 2.2 Behaviour of total cost with respect to inventory level and selling price in lockdown scenario 1.

Step 4: Execute step 5 to step 11 for all the values of 'i,' here 1 ≤ i ≤ length (S).
Step 5: Execute Step 6 to Step 11 for all values of 'j,' here 1 ≤ j ≤ length (L).
Step 6: Calculate the number of items to be stored in a warehouse L (j).
Step 7: Calculate the demand rate D(i, j).
Step 8: Calculate the inventory at any time t.
Step 9: Calculate $t_1(i,j)$ and $t_2(i,j)$.
Step 10: Calculate total costs.
Step 11: Search the selling price, inventory level to the minimum total cost.

2.3.2 Scenario 2

Similar to scenario 1, scenario 2 (Figure 2.3) is also studied in the time frame of 0 to t_2 amid a constant rate of deterioration. This scenario typically starts with a steady production P and demand rate D_O until it is disrupted due to enforcement of lockdown at time t_L. Due to the imposition of lockdown, the production P and demand D_O are disrupted by ΔP and Δd as shown in the time interval t_L to t_1. At time t_1, the production is stopped as the inventory reaches its maximum level. From t_1 to t_O the disrupted demand Δd continues as the lockdown period is still prevailing in this time interval. When the lockdown period is over as in the time interval t_O to t_2, the demand rate increases by Δd_1 due to panic purchasing outcome of the customers with a fear of unavailability of foods with a further extension in a lockdown situation. This situation continues until the warehouse inventory becomes zero.

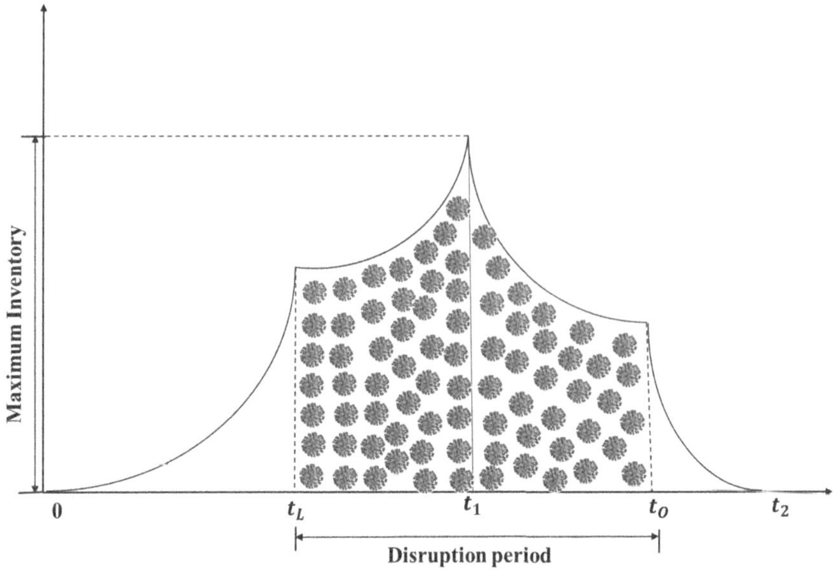

FIGURE 2.3 Production and dispatching of goods in lockdown scenario 2.

From 0 to t_L, inventory starts building up in a warehouse following the given differential equation:

$$\frac{dZ(t)}{dt} + \theta Z(t) = P - D_0 \quad (0 \le t \le t_L) \tag{2.15}$$

Taking the initial boundary condition at t = 0 and Z(t) = 0, equation (2.15) becomes:

$$Z(t) = \frac{P - D_0}{\theta}\left(1 - \frac{1}{e^{\theta t}}\right) \tag{2.16}$$

In the time interval t_L to t_1, the production and demand are disrupted by ΔP and Δd due to the imposition of lockdown. Therefore the governing equation will be:

$$\frac{dZ(t)}{dt} + \theta Z(t) = P - \Delta P - (D_0 - \Delta d) \quad (t_L \le t \le t_1) \tag{2.17}$$

Taking the boundary condition at $t = t_L$, equation (2.17) becomes:

$$Z(t) = \left(\frac{P - \Delta P - (D_0 - \Delta d)}{\theta}\right) + \left[(-\Delta P + \Delta d)\frac{e^{\theta t_L}}{\theta} - \frac{P - D_0}{\theta}\right]\frac{1}{e^{\theta t}} \tag{2.18}$$

At time $t = t_1$ and $Z(t_1) = L$, equation (2.18) becomes:

$$t_1 = \frac{1}{\theta} log\left\{\frac{(-\Delta P + \Delta d)\frac{e^{\theta t_L}}{\theta} - \left(\frac{P - D_0}{\theta}\right)}{L - \left(\frac{P - \Delta P - (D_0 - \Delta d)}{\theta}\right)}\right\} \tag{2.19}$$

At time $t = t_1$ to t_O, the production stops, but the lockdown still prevails in this period. The governing equation in this period is:

$$\frac{dZ(t)}{dt} + \theta Z(t) = -(D_0 - \Delta d) \quad (t_1 \le t \le t_O) \tag{2.20}$$

Using the boundary condition at $t = t_o$, equation (2.20) becomes:

$$Z(t) = -\left(\frac{D_0 - \Delta d}{\theta}\right) + \frac{e^{\theta t_1}\left(L + \left(\frac{D_0 - \Delta d}{\theta}\right)\right)}{e^{\theta t_0}} \tag{2.21}$$

From $t = t_O$ to t_2, lockdown periods end, and the government allows the movement of the people. Due to this, the demand increased by Δd_1, as mentioned earlier. The governing equation in this period is:

$$\frac{dZ(t)}{dt} + \theta Z(t) = -(D_0 + \Delta d_1) \quad (t_O \le t \le t_2) \tag{2.22}$$

Taking the boundary condition at $t = t_O$, equation (2.22) becomes:

$$Z(t) = -\left(\frac{D_0 + \Delta d_1}{\theta}\right) + \left\{\left(\frac{\Delta d + \Delta d_1}{\theta}\right)e^{\theta t_0} + e^{\theta t_1}\left(L + \left(\frac{D_0 - \Delta d}{\theta}\right)\right)\right\}\frac{1}{e^{\theta t}} \tag{2.23}$$

At $t = t_2$, $Z(t_2) = 0$, therefore:

$$t_2 = \frac{1}{\theta} log\left\{\frac{\left(\frac{\Delta d + \Delta d_1}{\theta}\right)e^{\theta t_0} + e^{\theta t_1}\left(L + \left(\frac{D_0 - \Delta d}{\theta}\right)\right)}{\frac{D_0 + \Delta d_1}{\theta}}\right\} \tag{2.24}$$

At $t = t_2$, $Z(t_2) = 0$ because the inventory is completely consumed until this period.

Now the value of variable costs is computed in the interval 0 to t_2.

(i) Holding cost, this includes all the expenses that arose due to holding the item in the inventory:

$$HC_{OW} = \left\{ H\int_0^{t_L} Z(t)\,dt + H\int_{t_L}^{t_1} Z(t)\,dt + H\int_{t_1}^{t_0} Z(t)\,dt + H\int_{t_{0'}}^{t_2} Z(t)\,dt \right\} \quad (2.25)$$

Suppose that: $HC_{OW} = \{L_9 + L_{10} + L_{11} + L_{12}\}$

$$L_9 = \int_0^{t_L} Z(t)\,dt = H\left(\frac{P - D_0}{\theta}\right)\left(t_L + \frac{1}{\theta}\left(e^{-\theta t_L} - 1\right)\right)$$

$$L_{10} = \int_{t_L}^{t_1} Z(t)\,dt = H\left\{\left[\frac{P - \Delta P - (D_0 - \Delta d)}{\theta}\right](t_1 - t_L) - \left[(D_0 - \Delta d)\frac{e^{\theta t_L}}{\theta} - \left(\frac{P - D_0}{\theta}\right)\right]\left(\frac{e^{-\theta t_1}}{\theta} - \frac{e^{-\theta t_L}}{\theta}\right)\right\}$$

$$L_{11} = \int_{t_1}^{t_0} Z(t)\,dt = H\left\{-\left(\frac{D_0 - \Delta d}{\theta}\right)(t_0 - t_1) - e^{\theta t_1}\left(L + \left(\frac{D_0 - \Delta d}{\theta}\right)\right)\left(\frac{e^{-\theta t_0}}{\theta} - \frac{e^{-\theta t_1}}{\theta}\right)\right\}$$

$$L_{12} = \int_{t_{0'}}^{t_2} Z(t)\,dt = H\left\{\left[\left(\frac{-(D_0 + \Delta d_1)}{\theta}\right)(t_2 - t_0)\right] - \left[\left(\frac{\Delta d + \Delta d_1}{\theta}\right)e^{\theta t_0} + e^{\theta t_1}\left(L + \left(\frac{D_0 - \Delta d}{\theta}\right)\right)\right]\left(\frac{e^{-\theta t_2}}{\theta} - \frac{e^{-\theta t_0}}{\theta}\right)\right\}$$

(ii) Deterioration cost, the cost experienced due to deterioration of items in the inventory:

$$DC = \theta c\int_0^{t_L} Z(t)\int dt + \theta c\int_{t_L}^{t_1} Z(t)\,dt + \theta c\int_{t_1}^{t_0} Z(t)\,dt + \theta c\int_{t_0}^{t_2} Z(t)\,dt \quad (2.26)$$

Suppose that: $DC = \{L_{13} + L_{14} + L_{15} + L_{16}\}$

$$L_{13} = \theta c\int_{0}^{t_L} Z(t)\,dt = c(P - D_0)\left(t_L + \left(\frac{e^{-\theta t_L}}{\theta} - \frac{1}{\theta}\right)\right)$$

$$L_{14} = \theta c\int_{t_L}^{t_1} Z(t)\,dt = \theta c\left\{\left(\frac{P - \Delta P - (D_0 - \Delta d)}{\theta}\right)(t_1 - t_L)\right.$$
$$\left. -\left[(\Delta P - \Delta d)\frac{e^{\theta t_L}}{\theta} - \left(\frac{P - D_0}{\theta}\right)\right]\left(\frac{e^{-\theta t_1}}{\theta} - \frac{e^{-\theta t_L}}{\theta}\right)\right\}$$

$$L_{15} = \theta c\int_{t_1}^{t_0} Z(t)\,dt = \theta c\left\{\left[-\left(\frac{D_0 - \Delta d}{\theta}\right)(t_0 - t_1)\right]\right.$$
$$\left. - e^{\theta t_1}\left(L + \left(\frac{D_0 - \Delta d}{\theta}\right)\right)\left(\frac{e^{-\theta t_0}}{\theta} - \frac{e^{-\theta t_1}}{\theta}\right)\right\}$$

$$L_{16} = \theta c\int_{t_0}^{t_2} Z(t)\,dt = \theta c\left\{-\left(\frac{D_0 - \Delta d_1}{\theta}\right)(t_2 - t_0) - \left[\left(\frac{\Delta d + \Delta d_1}{\theta}\right)e^{\theta t_0}\right.\right.$$
$$\left.\left. + e^{\theta t_1}\left(L + \left(\frac{D_0 - \Delta d}{\theta}\right)\right)\right]\left(\frac{e^{-\theta t_2}}{\theta} - \frac{e^{-\theta t_0}}{\theta}\right)\right\}$$

(iii) Purchase cost, the cost for purchasing the goods that are to be stored in the inventory:

$$PC = cL \tag{2.27}$$

All the relevant cost has been found out during the time interval 0 to t_2. Therefore, the total cost will be:

$$TC = HC_{ow} + DC + PC \tag{2.28}$$

Placing the values in equation (2.28), the total cost becomes:

$$TC(S,L) = H\left(\frac{P - D_0}{\theta}\right)\left(t_L + \frac{1}{\theta}\left(e^{-\theta t_L} - 1\right)\right) +$$
$$H\left\{\left[\frac{P - \Delta P - (D_0 - \Delta d)}{\theta}\right](t_1 - t_L) - \left[(D_0 - \Delta d)\frac{e^{\theta t_L}}{\theta} - \left(\frac{P - D_0}{\theta}\right)\right]\left(\frac{e^{-\theta t_1}}{\theta} - \frac{e^{-\theta t_L}}{\theta}\right)\right\}$$
$$+H\left\{-\left(\frac{D_0 - \Delta d}{\theta}\right)(t_0 - t_L) - e^{\theta t_1}\left(L + \left(\frac{D_0 - \Delta d}{\theta}\right)\right)\left(\frac{e^{-\theta t_0}}{\theta} - \frac{e^{-\theta t_1}}{\theta}\right)\right\} +$$
$$H\left\{\left[\left(\frac{-(D_0 + \Delta d_1)}{\theta}\right)(t_2 - t_0)\right] - \left[\left(\frac{\Delta d + \Delta d_1}{\theta}\right)e^{\theta t_0} + e^{\theta t_1}\left(L + \left(\frac{D_0 - \Delta d}{\theta}\right)\right)\right]\right.$$

$$\left(\frac{e^{-\theta t_2}}{\theta}-\frac{e^{-\theta t_0}}{\theta}\right)\Bigg\}+c(P-D_0)\left(t_L+\left(\frac{e^{-\theta t_L}}{\theta}-\frac{1}{\theta}\right)\right)+\theta c\left\{\left(\frac{P-\Delta P-(D_0-\Delta d)}{\theta}\right)\right.$$

$$(t_1-t_L)-\left[(\Delta P-\Delta d)\frac{e^{\theta t_L}}{\theta}-\left(\frac{P-D_0}{\theta}\right)\right]\left(\frac{e^{-\theta t_1}}{\theta}-\frac{e^{-\theta t_L}}{\theta}\right)\Bigg\}$$

$$+\theta c\left\{\left[-\left(\frac{D_0-\Delta d}{\theta}\right)(t_0-t_1)\right]-e^{\theta t_1}\left(L+\left(\frac{D_0-\Delta d}{\theta}\right)\right)\left(\frac{e^{-\theta t_0}}{\theta}-\frac{e^{-\theta t_1}}{\theta}\right)\right\}$$

$$+\theta c\left\{-\left(\frac{D_0-\Delta d_1}{\theta}\right)(t_2-t_0)-\left[\left(\frac{\Delta d+\Delta d_1}{\theta}\right)e^{\theta t_0}+e^{\theta t_1}\left(L+\left(\frac{D_0-\Delta d}{\theta}\right)\right)\right]\right.$$

$$\left(\frac{e^{-\theta t_2}}{\theta}-\frac{e^{-\theta t_0}}{\theta}\right)\Bigg\}+cL$$

The objective is to minimize the total cost. Therefore, the required condition for this is:

$$\frac{d^2TC(S,L)}{dS^2}\geq 0 \text{ and } \frac{d^2TC(S,L)}{dL^2}\geq 0$$

For simplicity, the convexity has been proved graphically using the Matlab software, as shown in Figure 2.4.

Algorithm for scenario 2: Same as that of scenario 1.

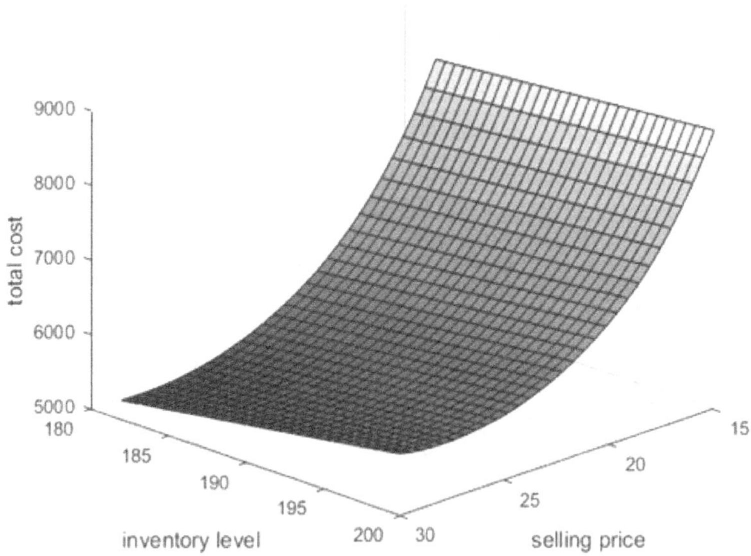

FIGURE 2.4 Behaviour of total cost with respect to inventory level and selling price in lockdown scenario 2.

2.4 NUMERICAL EXAMPLE

$$k = 100000, e = 2, \mathrm{P} = (1.5)\mathrm{D}_0, \Delta\mathrm{P} = \left(\frac{1}{3}\right)\mathrm{P}, \Delta\mathrm{P}_1 = \left(\frac{1}{2}\right)\mathrm{P}, \Delta\mathrm{d} = \left(\frac{1}{3}\right)\mathrm{D}_0,$$

$$\Delta\mathrm{d}_1 = \left(\frac{1}{2}\right)\mathrm{D}_0, \mathrm{H} = 1, \theta = 0.2, c = 12$$

$$For\,Scenario\,1, (t_L = 0.2, t_o = 0.5)$$

$$t_1 = 1.9, t_2 = 2.5, L = 182, S = 30, TC = 11126.8$$

$$For\,Scenario\,2, (t_L = 0.2, t_o = t_1 + 0.05)$$

$$t_1 = 3.1, t_2 = 3.6, L = 182, S = 29.48, TC = 5254.9$$

2.5 SENSITIVITY ANALYSIS

- It has been concluded from Table 2.2 that the total cost (TC) is heavily affected by the rate of production (P) in both scenarios. This is because an increase or decrease in production rate simultaneously changes the total cost.

TABLE 2.2
Behaviour of the Model w.r.t. Holding Cost (H), Deterioration Rate (θ), Production (P), Disrupted Production (ΔP), and Disrupted Demand (Δd)

		Scenario 1			Scenario 2		
		S	L	TC	S	L	TC
H	1	30	182	11,126.8	29.48	182	5254.9
	2	30	182	15,417.1	28.96	182	6053.1
	3	30	182	19,707.5	28.44	182	6848.7
	4	30	182	23,997.8	28.44	182	7642.3
θ	0.1	30	182	23,136.7	30	182	5052.1
	0.2	30	182	11,126.8	29.48	182	5260.4
	0.3	30	182	8346.8	25.34	182	6274.7
P	$(1.5)D_O$	30	182	11,126.8	29.48	182	5254.9
	$(1.6)D_O$	30	182	12,057.7	30	182	5248.5
	(1.7)Do	30	182	12,988.1	30	182	5344.0
	$(1.8)D_O$	30	182	13,917.9	30	182	5494.2
ΔP	(1/6)P	30	182	−12,771.4	30	182	4640.3
	(1/5)P	30	182	−4589.2	30	182	4704.1
	(1/4)P	30	182	6078.2	30	182	4835.0
	(1/3)P	30	182	11,126.8	29.48	182	5254.9

(Continued)

TABLE 2.2 Continued

		Scenario 1			Scenario 2		
		S	L	TC	S	L	TC
	(1/0.7)P	30	182	117,444.3	30	182	3932.8
	(1/0.5)P	30	182	213,482.7	30	182	3999.1
	(1/0.4)P	30	182	322,176.4	30	182	4026.3
Δd	(1/5)D_O	30	182	7508.7	24.31	182	6203.6
	(1/4)D_O	30	182	8760.9	26.37	182	5756.5
	(1/0.7)D_O	30	182	143,828.1	30	182	4318.4

TABLE 2.3
Model Analysis w.r.t. Selling Price (S) and Costs

	Scenario 1			Scenario 2		
S	HC	DC	TC	HC	DC	TC
15	11,561.2	12,712.6	26,457.9	734.94	5487.7	8406.6
20	7263.2	7966.5	17,413.7	647.06	3324.5	6278.3
25	5310.2	5799.0	13,293.2	778.45	2485.7	5448.1
30	4290.3	4652.5	11,126.8	808.27	2268.1	5260.4

- For a given cycle time in scenario 1, due to its shorter lockdown period, a disrupted production rate (ΔP) decreases the total production. Still, for a shorter interval of time, it is compensated soon by an increase in production (ΔP_1). Due to this, the total cost (TC) increases gradually.
- In scenario 2, the lockdown period is more extended; therefore, the total production rate is affected only by ΔP. Due to this, the entire production decreases, and total cost (TC) decreases.
- A disrupted demand is a function of the selling price, as mentioned earlier. Due to the shorter lockdown period in scenario 1, the disrupted demand (Δd) is simultaneously compensated by an increase in demand (Δd_1), resulting in a gradual increase in total cost. In scenario 2, the lockdown period is more significant because this disrupted demand (Δd) takes a longer time in compensating by demand (Δd1). Therefore, the total cost decreases for some time and after that increases gradually.
- In scenario 1, an increase in the deterioration rate decreases the total cost (TC) due to constant selling price. Conversely, in scenario 2, an increase in the deterioration rate increases the total cost as a decrease in unit selling price.
- From Table 2.3, it can be seen that an increase in selling price decreases the holding cost (HC) and deterioration cost (DC) in both scenarios; hence the total cost (TC) decreases.
- An increase in selling price increases the cycle time, as shown in Table 2.4.

TABLE 2.4
Model Analysis w.r.t. Selling Price (S) and Cycle Time

	Scenario 1			Scenario 2		
S	L	t_1	t_2	L	t_1	t_2
15	182	1.51	1.65	182	0.5	0.7
20	182	1.63	1.87	182	1.12	1.39
25	182	1.78	2.15	182	1.98	2.37
30	182	1.98	2.50	182	3.27	3.82

Managerial implications: This chapter gives some crucial managerial insights that will help mitigate the problems caused during this pandemic. These are as follows:

- The production rate should be managed efficiently. As the production rate increases, the inventory building up in the warehouse increases, increasing the inventory holding costs.
- The inventory should be built up in the warehouse as fast as possible. Disruption during the building-up phase may increase the total cost.
- The selling prices should be optimum, as the increase in selling price may help decrease the total cost, although it may result in customers' dissatisfaction.
- The cycle time increases with an increase in selling price. Therefore, to minimize the cycle time, the selling price should be optimized.

2.6 CONCLUSION, LIMITATIONS, AND FUTURE WORK

To examine the impact of the COVID-19 pandemic upon the warehouse system, the aforementioned mathematical model has been formulated for perishable inventory considering (i) disruptions in production and demand, (ii) different periods of lockdown, and (iii) price-sensitive demand. A numerical example has been shown to determine the effect of lockdown upon the total costs, along with their graphs and figures. The mathematical model helps determine the total cost incurred in both scenarios due to disruption. From this model, it has been concluded that the total cost of scenario 1 is more than scenario 2 in contrast with a shorter lockdown period. This is because scenario 1 takes longer to build up inventory than scenario 2 does. Due to this, the cycle time increased and hence the total cost increased. Therefore, it is advisable to build up the inventory in the warehouse as quickly as possible to avoid extra costs. All the factors that have been considered in this model are being properly synchronized with their costs. This model has become competent enough to provide all the necessary information such as the effect of production and demand disruptions, the length of the disruption period, the maximum level of inventory to be maintained in the warehouse, the selling price per unit item, and the effect of the deterioration rate.

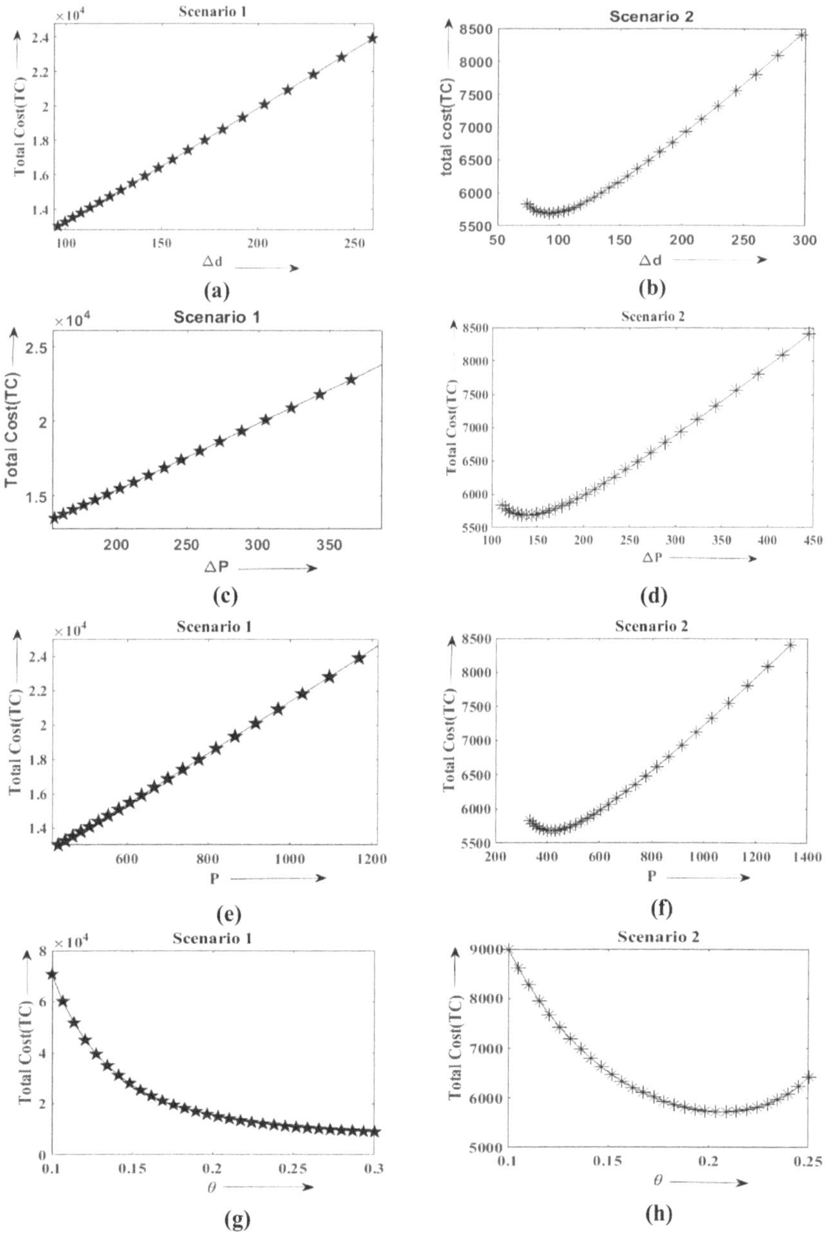

FIGURE 2.5 For lockdown scenarios 1 and 2; (a–b) total cost vs disrupted demand (Δd), (c–d) total cost vs disrupted production (ΔP), (e–f) total cost vs production (P), and (g–h) total cost vs deterioration rate (θ).

Though this model provides significant numbers of valuable information, it deals with certain limitations. These are as follows:

- It does not provide information related to inflation or the value of money over time.
- The effect of backlogging cannot be determined.
- It does not count the effects of two warehouse inventory policies on the total cost.

As the COVID-19 pandemic is not over to date, the previously given insights will support warehouse managers in framing out their future course. In addition, this study will help educate the distributors, wholesalers, and retailers in optimizing their inventories to sell them at optimum prices.

REFERENCES

Amjath-Babu, T. S., Krupnik, T. J., Thilsted, S. H., & McDonald, A. J. (2020). Key indicators for monitoring food system disruptions caused by the COVID-19 pandemic: Insights from Bangladesh towards effective response. *Food Security*, *12*(4), 761–768. https://doi.org/10.1007/s12571-020-01083-2

Barkur, G., & Kamath, G. B. (2020). Sentiment analysis of nationwide lockdown due to COVID 19 outbreak : Evidence from India. *Asian Journal of Psychiatry*, *51*(April), 102089. https://doi.org/10.1016/j.ajp.2020.102089

Basu, R. (2020). COVID control in India: A look back. *Journal of Comprehensive Health*, *8*(2), 129–131.

Baveja, A., Kapoor, A., & Melamed, B. (2020). Stopping Covid-19: A pandemic-management service value chain approach. *Annals of Operations Research*, *289*(2), 173–184. https://doi.org/10.1007/s10479-020-03635-3

Benos, L., Tsaopoulos, D., & Bochtis, D. (2020a). A review on ergonomics in agriculture. Part I: Manual operations. *Applied Sciences*. https://doi.org/10.3390/app10061905

Benos, L., Tsaopoulos, D., & Bochtis, D. (2020b). A review on ergonomics in agriculture. Part II: Mechanized operations. *Applied Sciences*. https://doi.org/10.3390/app10103484

Bhattacharya, A., Geraghty, J., Young, P., & Byrne, P. J. (2013). Design of a resilient shock absorber for disrupted supply chain networks: A shock-dampening fortification framework for mitigating excursion events. *Production Planning & Control*, *24*(8–9), 721–742. https://doi.org/10.1080/09537287.2012.666861

Bochtis, D., Benos, L., Lampridi, M., & Marinoudi, V. (2020). Agricultural workforce crisis in light of the COVID-19 pandemic. *Sustainability Article*, *2*. https://journalofcomprehensivehealth.co.in/jch/article/view/48

Brinca, P., Duarte, J. B., & Faria-e-Castro, M. (2020). Is the COVID-19 pandemic a supply or a demand shock? *Economic Synopses*, *31*. https://doi.org/10.20955/es.2020.31

Chiaramonti, D., & Maniatis, K. (2020). Security of supply, strategic storage and Covid19: Which lessons learnt for renewable and recycled carbon fuels, and their future role in decarbonizing transport? *Applied Energy*, *271*, 115216. https://doi.org/10.1016/j.apenergy.2020.115216

Christidis, P., & Christodoulou, A. (2020). The predictive capacity of air travel patterns during the global spread of the covid-19 pandemic: Risk, uncertainty and randomness. *International Journal of Environmental Research and Public Health*, *17*(10), 1–15. https://doi.org/10.3390/ijerph17103356

COVID-19 Pandemic – Impact on Food and Agriculture. (2019). Retrieved from http://www.fao.org/2019-ncov/q-and-a/impact-on-food-and-agriculture/en/

Deaton, B. J., & Deaton, B. J. (2020). Food security and Canada's agricultural system challenged by COVID-19. *Canadian Journal of Agricultural Economics*, *68*(2), 143–149. https://doi.org/10.1111/cjag.12227

Fortuna, G., & Foote, N. (2020). *Seasonal workers, CAP and COVID-19, farm to fork*. Retrieved from https://www.euractiv.com/section/agriculture-food/news/seasonal-workers-cap-and-covid-19-farm-to-fork/

Golan, M. S., Jernegan, L. H., & Linkov, I. (2020). Trends and applications of resilience analytics in supply chain modeling: Systematic literature review in the context of the COVID - 19 pandemic. *Environment Systems and Decisions*, *40*(2), 222–243. https://doi.org/10.1007/s10669-020-09777-w

Govindan, K., Mina, H., & Alavi, B. (2020). A decision support system for demand management in healthcare supply chains considering the epidemic outbreaks: A case study of coronavirus disease 2019 (COVID-19). *Transportation Research Part E: Logistics and Transportation Review*, *138*, 101967. https://doi.org/10.1016/j.tre.2020.101967

Gunessee, S., & Subramanian, N. (2020). Ambiguity and its coping mechanisms in supply chains lessons from the Covid-19 pandemic and natural disasters. *International Journal of Operations & Production Management*, *40*(7/8), 1201–1223. https://doi.org/10.1108/IJOPM-07-2019-0530

Hobbs, J. E. (2020). Food supply chains during the COVID-19 pandemic. *Canadian Journal of Agricultural Economics/Revue Canadienne d'agroeconomie*, *68*(2), 171–176. https://doi.org/10.1111/cjag.12237

Id, G. M. M. A., & Khatun, M. N. (2021). Impact of COVID-19 on vegetable supply chain and food security : Empirical evidence from Bangladesh. *PLoS ONE*, *16*(3), 1–12. https://doi.org/10.1371/journal.pone.0248120

Ivanov, D. (2020). Predicting the impacts of epidemic outbreaks on global supply chains: A simulation-based analysis on the coronavirus outbreak (COVID-19/SARS-CoV-2) case. *Transportation Research Part E: Logistics and Transportation Review*, *136*(March). https://doi.org/10.1016/j.tre.2020.101922

Jabbarzadeh, A., Fahimnia, B., Sheu, J.-B., & Moghadam, H. S. (2016). Designing a supply chain resilient to major disruptions and supply/demand interruptions. *Transportation Research Part B: Methodological*, *94*, 121–149. https://doi.org/10.1016/j.trb.2016.09.004

Jámbor, A., Czine, P., & Balogh, P. (2020). The impact of the coronavirus on agriculture: First evidence based on global newspapers. *Sustainability*. https://doi.org/10.3390/su12114535

Leite, H., Lindsay, C., & Kumar, M. (2021). COVID-19 outbreak: Implications on healthcare operations. *The TQM Journal*, *33*(1), 247–256. https://doi.org/10.1108/TQM-05-2020-0111

Majumdar, A., Shaw, M., & Sinha, S. K. (2020). COVID-19 debunks the myth of socially sustainable supply chain: A case of the clothing industry in South Asian countries. *Sustainable Production and Consumption*, *24*, 150–155. https://doi.org/10.1016/j.spc.2020.07.001

Mor, R. S., Srivastava, P. P., Jain, R., Varshney, S., & Goyal, V. (2020). Managing food supply chains post COVID-19: A perspective. *International Journal of Supply and Operations Management*, *7*(3), 295–298. https://doi.org/10.22034/IJSOM.2020.3.7

Paul, S. K., & Chowdhury, P. (2020). A production recovery plan in manufacturing supply chains for a high-demand item during COVID-19. *International Journal of Physical Distribution and Logistics Management*, ahead-of-print. https://doi.org/10.1108/IJPDLM-04-2020-0127

Paul, S. K., & Chowdhury, P. (2021). A production recovery plan in manufacturing supply chains for a high-demand item during COVID-19. *International Journal of Physical Distribution & Logistics Management*, *51*(2), 104–125. https://doi.org/10.1108/IJPDLM-04-2020-0127

Rana, R. S., Kumar, D., Mor, R. S., & Prasad, K. (2021). Modelling the impact of demand disruptions on two warehouse perishable inventory poli+cy amid COVID-19 lockdown. *International Journal of Logistics Research and Applications*, 1–23. https://doi.org/10.1080/13675567.2021.1892043

Rana, R. S., Kumar, D., & Prasad, K. (2020). Two warehouse dispatching policies for perishable items with freshness efforts, inflationary conditions and partial backlogging. *Operations Management Research*, Ahead of Press. https://doi.org/10.1007/s12063-020-00168-7

Remko, van H. (2020). Research opportunities for a more resilient post-COVID-19 supply chain – closing the gap between research findings and industry practice. *International Journal of Operations & Production Management*, *40*(4), 341–355. https://doi.org/10.1108/IJOPM-03-2020-0165

Richards, T. J., & Rickard, B. (2020). COVID-19 impact on fruit and vegetable markets. *Canadian Journal of Agricultural Economics*, *68*(2), 189–194. https://doi.org/10.1111/cjag.12231

Sarkar, S., & Kumar, S. (2015). A behavioural experiment on inventory management with supply chain disruption. *International Journal of Production Economics*, *169*, 169–178. https://doi.org/10.1016/j.ijpe.2015.07.032

Toffolutti, V., Stuckler, D., & McKee, M. (2020). Is the COVID-19 pandemic turning into a European food crisis? *European Journal of Public Health*, *30*(4), 626–627. https://doi.org/10.1093/eurpub/ckaa101

Vancic, A., & Pärson, G. F. A. (2020). *Changed buying behaviour in the COVID-19 pandemic: The influence of price sensitivity and perceived quality*. Kristianstad University, SE-291 88 Kristianstad, Sweden, +46442503000, Retrieved from www.hkr.se

WTO. (2020). *Trade set to plunge as COVID-19 pandemic upends global economy*. Retrieved from https://www.wto.org/english/news_e/pres20_e/pr855_e.htm

Yuen, K. F., Wang, X., Ma, F., & Li, K. X. (2020). The psychological causes of panic buying following a health crisis. *International Journal of Environmental Research and Public Health*. https://doi.org/10.3390/ijerph17103513

3 Sustainable Supply Chain Resilience

A Decision Framework to Manage Disruptions and Retain Sustainability

Varun Sharma and Bijaya K. Mangaraj

CONTENTS

3.1 INTRODUCTION

The COVID-19 pandemic has disrupted the supply chain operations of 63% of businesses and caused a delay of more than four weeks for more than 50% of the companies in India (FICCI, 2020). The supply of the products is severely impacted as the

DOI: 10.1201/9781003150084-3

top 1000 companies have more than 12,000 manufacturing facilities in COVID-19 quarantine areas. Once hit by such events, very few supply chains that are resilient enough would be able to get back to their earlier operations completely (Reeves et al., 2020). Resilient strategy for supply chains provides the perturbations' absorption capacity from external disruptions and helps businesses recover quickly to their regular operations as soon as possible (Handmer & Dovers, 1996; Sheffi et al., 2003). However, the adoption of only a resilience-based strategy will not ensure the continuity of operations. The recent orientation of government bodies and consumer preferences has been increasingly inclining towards sustainability. Therefore, firms are under pressure from their stakeholders to adopt sustainability to ensure their business longevity (Deng et al., 2019; Seuring & Müller, 2008). Through sustainability, supply chains can mitigate the risks originating from the environment and society more effectively (Shad et al., 2019; Valinejad & Rahmani, 2018). Therefore, adopting a resilience strategy while preserving the supply chain's sustainability objectives may improve the supply chain's recovery from major disruptions.

According to the literature, the relationship between supply chain resilience (SCR) and sustainable supply chain (SSC) is not always complementary. There is also evidence suggesting the possibility of a contradictory relationship between them (Karutz et al., 2018; Marchese et al., 2018). Thus, managers face challenges in operationalizing the synergy of SCR and SSC. Therefore, this study uses a paradox perspective to deal with the dilemma of adopting SCR and SSC simultaneously. The paradox perspective enables us to achieve a symbiosis between two competing alternatives by emphasizing the adoption of the commonalities between them (Smith & Lewis, 2011). There could be two possibilities of integrating SCR and SSC; first, enhancing SSC by incorporating resilience. Second, strengthen SCR through sustainability. The literature includes papers on enhancing SSC through SCR, but no studies attempted to improve SCR through SSC (Fiksel, 2006; He et al., 2020). The available literature also indicates that there are no studies that facilitate the SCR's operationalization by listing out the relevant practices. Additionally, the techniques used in such literature also neglect the behavioural aspects of the decision makers. Therefore, this study formulates the following research objectives to address the gaps in the literature.

a. To develop a human-centric decision-making framework that enables the integration of SCR and SSC.
b. To identify a list of relevant practices that enhance SCR while fulfilling sustainability objectives.

The study identifies a comprehensive list of practices through an in-depth literature review on SCR, followed by an evaluation based on the three sustainability dimensions. The evaluation is conducted through the proposed method that manages the subjectivity, vagueness, and irrationalities of data during the decision-making process. The proposed methodology is validated in the context of an Indian automobile industry. Managers adopting the proposed method can achieve sustainable supply chain resilience (SSCR), in which resilience is enhanced through incorporating sustainability goals.

3.2 LITERATURE REVIEW

3.2.1 Supply Chain Resilience and Sustainable Supply Chain

The concept of supply chain resilience has been preferred by many researchers to handle such disruptions and ensure quick recovery of the supply chain operations (Adobor & McMullen, 2018; Aven, 2018; Christopher & Peck, 2004; Rice & Sheffi, 2005). SCR is defined as "the adaptive capability of the supply chain to prepare for unexpected events, respond to disruptions, and recover from them by maintaining continuity of operations at the desired level of connectedness and control over structure and function" (Ponomarov & Holcomb, 2009). These capabilities also strengthen the firm's overall structure and aid in managing a great range of risks. However, the recent literature signals that resilience is not enough to deal with the risks associated with climate change and social structure in a supply chain (Ghadge et al., 2020; Irvin, 2020). Therefore, many researchers have advocated adopting a sustainable supply chain (SSC) management strategy to deal specifically with environmental and social risks (Giannakis & Papadopoulos, 2016; Hofmann et al., 2014; Roehrich et al., 2014). An SSC helps manage the risks that originated due to the impact of supply chain operations on the environment and society. At the same time, SCR will ensure the preparedness of a firm against any unexpected disturbances from its surroundings.

The SSC is based on the idea of sustainable development that ensures intergenerational equality of available resources (WCED, 1987). Since the produced goods flow through a supply chain and pose a burden on its social and environmental surroundings, it becomes necessary to integrate sustainability with the supply chains (Ageron et al., 2012; The Sustainability Consortium, 2016). The SSC is defined as "the strategic, transparent integration and achievement of an organization's social, environmental, and economic goals in the systemic coordination of key inter-organizational business processes for improving the long-term economic performance of the individual company and its supply chains." According to the SSC, the supply chain's performance goals are to be set on the triple bottom line, where social, environmental, and economic goals are given equal importance (Govindan et al., 2013; Seuring & Müller, 2008).

Thus, SCR and SSC are two strategies that need to be adopted simultaneously to deal with unexpected disruptions and sustainability-related risks (Achour et al., 2015; Anderies et al., 2013; Shamout et al., 2021). Furthermore, the literature suggests that adopting the two strategies simultaneously is crucial to operating under high competitiveness and turbulence (Anderies et al., 2013; Jabbarzadeh et al., 2018; Rajesh, 2018a; Ramezankhani et al., 2018; Bellamy et al., 2019). However, the objective of SSC is to increase the efficiency of the supply chain, which makes supply chains more vulnerable, reducing their resilience (Fahimnia & Jabbarzadeh, 2016; Scholten & Schilder, 2015). On the other hand, SCR's objective is to recover quickly, which requires redundant resources, hence reducing the sustainability of the supply chain (Barroso et al., 2015). Thus, SCR and SSC share an overlapping relationship with many contradictions and dependencies between them.

SCR's primary objective is to deal with disruptions in supply chain operations. However, SSC could focus on the disruptions that could arise due to only environmental and social impacts (Abdel-Basset & Mohamed, 2020; Giannakis & Papadopoulos, 2016; Shad et al., 2019; Wijethilake & Lama, 2019). The managers are striving for better disruption management strategies, but due to continuous overlooking of environment-related risks, they cannot achieve a higher level of SCR (Mohammed et al., 2019). Also, due to SCR's and SSC's contradictory objectives, there is a possibility that implementation of SCR practices could lead to lower SSC performance, which can further escalate the sustainability-related risks (Karutz et al., 2018; Marchese et al., 2018). For example, specific SCR measures such as large supplier base, stockpiling, and extra capacity may violate the sustainability objectives of resource efficiency, relationship with the local community, and low carbon footprint, respectively. Thus, SSC and SCR share a mixed complementary and contradictory relationship at various intersections. Therefore, there is a need to integrate the two by reducing trade-offs between them and enhancing supporting elements to achieve synergy.

3.2.2 Paradoxical Lens to Integrate SCR and SSC

The aforementioned literature highlights a mix of contradictory and contributory relationships SCR and SSC share. The integration of the two strategies can be best studied through the paradoxical lens (Smith & Lewis, 2011). According to Smith and Lewis (2011), a paradoxical lens facilitates the continuous management of competing demands like SCR and SSC in an organizational system. The paradox theory suggests that the paradoxes due to competing demands are inherent in organizations and can be initiated owing to surrounding conditions, individual sense-making, or relational dialogues (Smith & Tracey, 2016). Stability–flexibility, commitment–change, and established routines–novel approaches are examples of the paradoxes in an organizational system (Smith, 2014). The theory proposes to own both sides of the paradox by avoiding the contradictions and accepting the complementary opportunities at the intersection of the two competing approaches. It facilitates the paradoxical resolution between the two divergent strategies to enable maximum utilization of their potentials (Smith & Lewis, 2011). Therefore, the paradox lens could provide a suitable perspective while studying the relationship between SCR and SSC. Since SCR has several practices and SSC has three-dimensional goals, a decision-making framework can evaluate the SCR practices based on their ability to achieve SSC goals. The evaluation will give a clear distinction between high- and low-performing SCR practices. Adopting high-performing practices will help managers to integrate SSC into SCR.

3.2.3 Decision-Making Techniques

In recent literature, several authors have realized the need to integrate resilience and sustainability strategies. They have attempted to obtain the synergy of the two strategies using various decision-making techniques. According to the literature, the decision-making techniques used by researchers in the context of SCR and SSC mainly comprise objective optimization techniques (Fahimnia et al., 2018; Hosseini-Motlagh et al., 2020; Jabbarzadeh et al., 2018; Mousavi Ahranjani et al., 2020; Zahiri et al., 2017; Zamanian et al., 2020). The recent literature makes it evident that there is an increase in the studies comprising multi-attribute decision-making techniques (Ghadge et al., 2020; He et al., 2020; Kaur et al., 2020; Kayikci, 2020; Mohammed, 2020). The techniques mentioned in the literature like VIKOR (VIekriterijumsko Kompromisno Rangiranje), TOPSIS (Technique for Order of Preference by Similarity to Ideal Solution), and DEMATEL (Decision Making Trial and Evaluation Laboratory), which are frequently used, assume that the decision maker always acts rationally. Whereas in real situations, decision makers can sometimes act irrationally and make skewed decisions. Moreover, the data collected from the experts can be vague due to the uncertainty of human assessment (Kayikci, 2020; Mousavi Ahranjani et al., 2020; Zamanian et al., 2020). Most studies have taken care of the data's vagueness in literature by using the fuzzy theory in their analysis. However, there is no evidence of any other technique such as grey theory (see Section 3.3.3.2) to handle the uncertainties appearing due to the decision maker's judgmental decisions. Therefore, this paper attempts to integrate the SCR and SSC strategies through a paradoxical lens using decision-making techniques that take a human-centric decision-making approach.

3.3 PROPOSED RESEARCH METHODOLOGY

The proposed framework in this paper follows a two-step process (see Figure 3.1). Initially, there is an identification of the supply chain resilience practices and suitable sustainability criteria through an extant literature review followed by recording industry experts' opinions. Then, the opinions are analyzed using a combination of decision-making techniques.

3.3.1 IDENTIFYING SUPPLY CHAIN PRACTICES

For this study, the major supply chain resilience enablers are listed from various sources. A total of eight enablers of supply chain resilience are identified. The

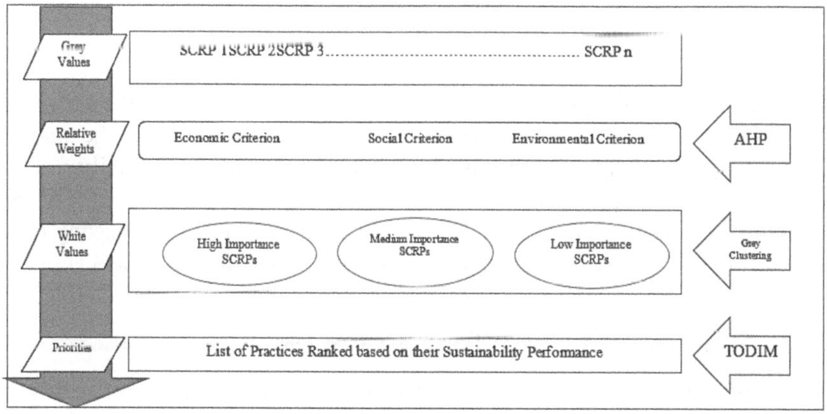

FIGURE 3.1 Proposed research framework.

following section states the meaning of the enablers and how they contribute towards resilience in the supply chain:

1. Agility: The alertness of the supply chain towards internal and external changes. It helps to enhance the capability of the supply chain to respond in a timely and flexible manner (Li et al., 2009).
2. Collaboration: The joint planning, management, execution, and performance action by two or more companies coming together (Anthony, 2000). It acts as a cohesive force that brings the companies together during a given crisis and helps them recover (Autry & Glenn Richey, 2009).
3. Flexibility: The degree to which the supply chain of a firm can adjust its speed, destinations, and volumes as per the requirement of the market (Lummus et al., 2005).
4. Risk management culture: Identifies and eliminates the sources of risk through mitigation strategies among supply chain managers (Manuj & Mentzer, 2008).
5. Visibility: Refers to the ability to be seen by supply chain partners. It is decided based on the access to critical information by supply chain partners, which helps them make crucial business decisions (Barratt, 2004).
6. Transparency: A way of making supply chain commitments more meaningful through sharing information about suppliers and supply chain processes with all the stakeholders. It facilitates crucial information transfer from the firm to its stakeholders (Egels-Zandén et al., 2015).
7. Redundancy: To achieve redundancy, firms need to keep their resources more than required to use these additional resources to overcome the situation in case of any contingency. In a supply chain, maintaining safety stock, having multiple suppliers, and running operations at low-capacity utilization rates are some of the practices followed to create redundancy (Blackhurst et al., 2005).
8. Anticipation capability: Refers to the ability of a firm to accurately predict future events of change so that it can prepare itself for any adverse situation. This capability of the firm is an integral part of a proactive strategy for resilience (Wieland & Marcus Wallenburg, 2013).

The study identifies the practices associated with the enablers from the existing literature. However, some of the enablers are interrelated, as are some of the practices. Due to the overlapping relationship of the enablers, there is a possibility of having redundant practices. Therefore, the list of practices was sent to a panel of three experts to identify the redundant practices. After identifying and removing redundant practices, a clear set of 31 resilient practices is obtained (see Table 3.1).

3.3.2 Sustainability Criteria: Triple-Bottom-Line

The identification of suitable criteria is made under three dimensions of sustainability, that is, economic, social, and environmental. It is done to prioritize the practices as per their ability to fulfil sustainability goals.

TABLE 3.1
List of Supply Chain Resilient Practices After Removing Redundancies

SCRP No.	Practices
SCRP 1	Reduce manufacturing lead time
SCRP 2	Reduce development cycle time
SCRP 3	Increase frequencies of new product introductions
SCRP 4	Increase levels of product customization
SCRP 5	Improve delivery capability
SCRP 6	Improve customer service
SCRP 7	Improve delivery reliability
SCRP 8	Improve responsiveness to changing market needs
SCRP 9	Promote joint planning
SCRP 10	Adopt joint problem solving
SCRP 11	Adopt joint performance measurement
SCRP 12	Leverage resources and skills of partners
SCRP 13	Increase the ability to change the quantity of the supplier's order
SCRP 14	Reduce delivery times of supplier's order
SCRP 15	Increase ability to alter delivery schedules to meet customer requirement
SCRP 16	Increase ability to change production volume capacity
SCRP 17	Accommodate changes in production mix
SCRP 18	Avoid risk
SCRP 19	Use vertical integration
SCRP 20	Stockpile
SCRP 21	Use contracts
SCRP 22	Joint efforts in continuity plans share information and increase visibility
SCRP 23	Develop multiple and local sourcing
SCRP 24	Maintain safety stock
SCRP 25	Develop backup sites
SCRP 26	Maintain excess capacity
SCRP 27	Include names of the firm's supplier's corporate disclosure
SCRP 28	Include information about the sustainability conditions of the supplier's corporate disclosure
SCRP 29	Include information on the purchasing practices of the buying firms
SCRP 30	Develop the ability to sense information about external events and changes in the supply chain
SCRP 31	Develop the ability to obtain more complete information to support decision-making for managing different kinds of dependency in supply chain relationships

3.3.2.1 Economic Performance: Resource Management Efficiency (RME)

The variables that indicate the flow of money are used to measure a company's economic performance (Slaper & Hall, 2011). This is a traditional criterion of performance that gives importance to any business process's financial outcomes (John, 1997). There can be two ways to increase economic performance: either increasing revenue or decreasing the associated cost to any product. The significant cost price of any product includes procuring, managing, and operating the resources in a supply chain. Raw materials, final products, semi-finished products, machinery, packaging materials, and human resources are examples of the resources (Zhou et al., 2000). If the supply chain strategy manages the resources more efficiently, it will directly reduce the costs to achieve financial benefits for the overall supply chain (Harmon & Cowan, 2009; Zailani et al., 2012). Hence, resource management efficiency is considered as an essential criterion for measuring economic performance in the supply chain.

3.3.2.2 Social Performance: Relationship with Community Stakeholders (RCS)

Social performance represents the people's aspect as a part of the 3Ps concept (People, Planet, and Profit) (Elkington, 1997), which means that the firms must be working towards developing their stakeholders for better social performance. There can be three types of stakeholders in a supply chain – internal, inter-firm, and external (Klassen & Vereecke, 2012). Under risk management activities, it is vital to study the stakeholders and manage them efficiently. This results in maximizing their positive influence and reducing their negative influence, as they have a considerable impact on the business (Walker & Bourne, 2007). Local communities have been a critical influencer as external stakeholders and are among the most challenging stakeholders to manage (Stephen Tsang et al., 2009; Walker & Bourne, 2007). Handling the challenges of managing the local community will result in a better relationship between the firm and the community. Owing to the same, we have considered it as an essential criterion for assessing the social performance of the firm.

3.3.2.3 Environmental Performance: Reduced Energy Consumption (REC)

Firms that can reduce the impact of their operations on the surrounding natural environment are considered environmentally sustainable companies (Eltayeb et al., 2011). In the case of a supply chain, the environment is affected by the carbon footprint, as the operations are dependent on the energy produced from fossil fuels (Thuermer, 2008). To reduce the impact on the environment, firms need to either reduce energy consumption or look for alternative sustainable energy sources other than fossil fuels (Carter & Rogers, 2008). In the context of logistics and supply chains, energy efficiency is usually neglected, but at the same time, it has a significant impact on the environment. Hence to improve the environmental performance in the supply chain, there is a need to look for options to reduce energy consumption (Kovács & Halldórsson, 2010). In this study, we have taken into account reduced energy consumption as the criterion for environmental performance.

3.3.3 Integrated AHP-Grey-TODIM

According to Section 3.2.3, popular multi-criteria decision-making (MCDM) techniques used for SCR and SSC do not incorporate irrationality, vagueness, and subjectivity of the decision makers' opinions. Therefore, this study attempts to fill this gap by proposing an integrated decision framework that deals with human behavioural factors. These factors can influence the decision-making process if appropriate measures are not deployed. Thus, this paper integrates three MCDM techniques that consider a human-centric approach to achieve the synergy between SCR and SSC.

TABLE 3.2
Experts' Profile

Experts	Designation	Experience (in years)	Role
Expert 1	Deputy General Manager	23	Supplier Development
Expert 2	Deputy Manager	6	Supply Planning
Expert 3	Assistant Manager	6	Supplier Support
Expert 4	Deputy General Manager	11	Supply Planning
Expert 5	Deputy Manager	8	Supplier Quality Assurance

The following section validates the previously integrated decision framework by demonstrating an example of XYZ automobile company. The company based in India has realized that its supply chain operates in highly turbulent surroundings due to the COVID-19 pandemic situation across the world. Consequently, the managers decide to adopt the SCR strategy to ensure quick recovery of the operations. They are also facing continuous pressure from their shareholders, customers, and government to embrace sustainability in their operations. However, they know that adopting resilience can have a mixed impact on the sustainability objectives of the supply chain. Subsequently, they also realize the role of SSC in managing risk from social and environmental events. Thus, they decide to list certain supply chain resilient practices (SCRPs) that their firm and its suppliers can adopt. These practices should essentially help them to achieve SSCR. Therefore, the managers plan to evaluate the SCRPs based on their ability to achieve sustainability goals and adopt only those that contribute to the sustainability goals. The proposed framework helps the managers in the aforementioned decision-making process (see Figure 3.1).

The study uses the opinions from the supply chain working professionals with at least five years of experience at the managerial level or above in the automobile industry. It uses the purposive sampling technique to ensure that the selected experts have significant knowledge regarding the automobile supply chain and its sustainability goals. Since the higher number of experts can cause inconsistencies in the analysis (Pun & Hui, 2001; Tarei et al., 2018), the study uses five experts. The profile of the selected experts is mentioned in Table 3.2.

3.3.3.1 Analytical Hierarchy Process (AHP)

The AHP technique manages the tangible and non-tangible objectives of the problem at hand. It successfully deals with the individuals' subjective ideas and beliefs about themselves (R. W. Saaty, 1987; T. L. Saaty, 2008; Suresh & Kaparthi, 1992). The AHP involves a decision-making algorithm that uses judgments based on a pairwise comparison process between different criteria. The technique gives normalized weights and ranks to a set of criteria by using experts' subjective judgments. In this study, the sustainability criteria for the problem carry different and relative importance. This relative importance varies as per the perception of the decision makers.

The identified experts were asked to give their preferences for the aforementioned criteria in a pairwise comparison manner. The data was collected on a scale of 1 to 9, where 1 means equally preferred and 9 means extremely preferred. Then the weights of the criteria are calculated using the following steps for the AHP.

Step 1: Getting the preference matrices from experts
Step 2: Structure of the judgment matrix

Pairwise comparison data recorded from experts is organized in the form of matrices. As there are three criteria, each expert corresponds to a 3×3 judgment matrix.

Step 3: Normalization of the judgment matrix
Step 4: Weight calculation from the judgment matrix

The row-wise average of the normalized matrix elements gives us the weights of the criteria relative to the expert.

Step 5: Checking the consistency of the judgment matrix

For calculating the consistency ratio of the judgment matrix, the column weighted-sum of the matrix, obtained after step 2, is calculated. Dividing the weighted sum value by the allocated weights, eigenvalues are calculated, and then an average of all the three will give the value of the maximum eigenvalue (λ_{max}). Using the λ_{max}, the consistency index (CI) is calculated using equation (3.1) to check the inconsistency in the judgment matrix., where *n* is the number of criteria.

$$CI = \frac{\lambda_{max} - n}{n - 1} \tag{3.1}$$

The consistency ratio (CR) is calculated as CI/RI, where RI is the random consistency index and depends on the total number of criteria. The values of RI for different criteria are adopted from the consistency index table available in Wind and Saaty (1980). The value of the consistency ratio must be smaller than 0.1. If the CR value exceeds 0.1, then the judgments will not be reliable due to randomness in experts' responses (T. L. Saaty, 2008).

Step 6: Weights of the sustainability criteria

Suppose the consistency ratio of all the matrices corresponding to all the experts is lower than 0.1. In that case, the study moves ahead by taking an average of the weights for the different experts' criteria.

Table 3.3 presents the calculated weights from the responses of all five experts. The consistency ratio for each expert was also calculated and found to be below 0.10. The final weights are calculated by taking an arithmetic mean of the weights obtained from the five experts. Thus, the RME, REC, and RCS weights are 0.5732, 0.2336, and 0.1926, respectively.

TABLE 3.3
AHP Responses and Criteria Weights

Experts	Criteria	RME	RCS	REC	Criteria Weights	Consistency Ratio
Expert 1	RME	1	3	4	61.40%	0.077
	RCS	1/3	1	3	26.80%	
	REC	1/4	1/3	1	11.70%	
Expert 2	RME	1	2	5	59.50%	0.006
	RCS	1/2	1	2	27.60%	
	REC	1/5	1/2	1	12.80%	
Expert 3	RME	1	5	3	63.70%	0.04
	RCS	1/5	1	1/3	10.50%	
	REC	1/3	3	1	25.80%	
Expert 4	RME	1	1	1	33.30%	0.098
	RCS	1	1	1	33.30%	
	REC	1	1	1	33.30%	
Expert 5	RME	1	5	4	68.70%	0
	RCS	1/5	1	2	18.60%	
	REC	1/4	1/2	1	12.70%	
		Average Weights		RME	**57.32%**	
				RCS	**23.36%**	
				REC	**19.26%**	

3.3.3.2 Grey Clustering

The study uses the grey clustering technique to categorize the practices into smaller clusters based on their level of importance for achieving sustainability. This method successfully deals with the managers' confusion regarding a long list of interrelated practices. It uses a set of unsure data and converts them into more valuable data through the whitenization process (Lin et al., 2009; Wen, 2008). It does not require strict statistical procedures and inference rules and helps achieve classification within unsure or judgmental data (Rajesh, 2018b).

In this study, there is a list of 31 supply chain resilient practices selected from the literature. All the practices are not entirely independent of each other and can have a few overlaps. It is also observed that the list has become highly extensive due to several interrelated practices. This can create confusion for the supply chain managers during the implementation process. Therefore, it is convenient and more comfortable for them to focus on a smaller set of practices than a large one. The grey clustering technique classifies the SCRPs based on their level of importance through grey classes: high, medium, and low importance.

The supply chain experts were asked about the impact of 31 SCRPs on the three sustainability goals. Their responses were collected using linguistic variables mentioned in Table 3.4. The median method is used to find out the average of the five responses. The linguistic variables are converted to their corresponding grey values (see Table 3.4). Finally, the grey values corresponding to experts' opinion is represented in Table 3.5. The grey value of impact of ith SCRP over j^{th} sustainability goal is represented by $\otimes G_{ij}$.

TABLE 3.4
Linguistic Variables and Their Corresponding Grey Values

Linguistic Values	Grey Values
Extremely Low (EL)	(1,2)
Very Low (VL)	(2,3)
Low (L)	(3,4)
Moderately Low (ML)	(4,5)
Medium (M)	(5,6)
Moderately High (MH)	(6,7)
High (H)	(7,8)
Very High (VH)	(8,9)
Extremely High (EH)	(9,10)

TABLE 3.5
Grey Values Corresponding to SCRPs' Evaluation

Weights	0.5732		0.2336		0.1926	
S. No.	RME		REC		RCS	
1	8	9	6	7	6	7
2	7	8	6	7	7	8
3	4	5	7	8	4	5
4	8	9	7	8	3	4
5	7	8	8	9	5	6
6	7	8	7	8	4	5
7	6	7	7	8	4	5
8	7	8	7	8	5	6
9	8	9	7	8	7	8
10	8	9	8	9	7	8
11	7	8	5	6	6	7
12	8	9	8	9	7	8
13	7	8	7	8	6	7
14	7	8	6	7	5	6
15	6	7	7	8	5	6
16	7	8	5	6	7	8
17	4	5	7	8	4	5
18	6	7	4	5	3	4
19	7	8	6	7	6	7
20	3	4	5	6	3	4
21	6	7	6	7	4	5
22	7	8	7	8	7	8
23	7	8	5	6	6	7
24	5	6	5	6	3	4

Weights	**0.5732**		**0.2336**		**0.1926**	
S. No.	**RME**		**REC**		**RCS**	
25	5	6	4	5	3	4
26	2	3	3	4	3	4
27	3	4	5	6	3	4
28	4	5	6	7	6	7
29	7	8	7	8	4	5
30	7	8	6	7	4	5
31	7	8	6	7	5	6

Weighted grey values for each of the practices are calculated by multiplying the weights of the criteria with their grey values ($\otimes\ddot{G}_{ij} = w_j * \otimes G_{ij}$) in Table 3.5. In Table 3.6, the weighted grey values of the impact of each practice for all three criteria are obtained. Then, the grey-weighted values are converted into crisp values using three steps.

I. Normalizing grey values using equations (3.2), (3.3), and (3.4):

$$\overline{\otimes}\widetilde{G}_{ij} = \left(\overline{\otimes}\ddot{G}_{ij} - {}^{min}_{j}\overline{\otimes}\ddot{G}_{ij}\right) / \Delta_{min}^{max} \tag{3.2}$$

$$\underline{\otimes}\widetilde{G}_{ij} = \left(\underline{\otimes}\ddot{G}_{ij} - {}^{min}_{j}\underline{\otimes}\ddot{G}_{ij}\right) / \Delta_{min}^{max} \tag{3.3}$$

$$\Delta_{min}^{max} = {}^{max}_{j}\overline{\otimes}\ddot{G}_{ij} - {}^{min}_{j}\underline{\otimes}\ddot{G}_{ij} \tag{3.4}$$

Where $\overline{\otimes}$ is upper, $\underline{\otimes}$ is lower and $\widetilde{G}_{ij}$ is normalized values of the grey number $\otimes\ddot{G}_{ij}$ respectively:

II. Converting the normalized grey values into normalized white values using equation (3.5);

$$W_{ij} = \left(\frac{\left(\underline{\otimes}\widetilde{G}_{ij}\left(1-\underline{\otimes}\widetilde{G}_{ij}\right)\right)+\left(\overline{\otimes}\widetilde{G}_{ij}\right)^{2}}{1-\underline{\otimes}\widetilde{G}_{ij}+\overline{\otimes}\widetilde{G}_{ij}}\right) \tag{3.5}$$

III. Calculating the final white values (Table 3.7) using equation (3.6):

$$W_{ij}^{*} = \left({}^{min}_{j}\underline{\otimes}\ddot{G}_{ij} + W_{ij} * \Delta_{min}^{max}\right) \tag{3.6}$$

TABLE 3.6
Weighted Grey Values

S. No.	RME		REC		RCS	
1	4.5856	5.1588	1.4016	1.6352	1.1556	1.3482
2	4.0124	4.5856	1.4016	1.6352	1.3482	1.5408
3	2.2928	2.866	1.6352	1.8688	0.7704	0.963
4	4.5856	5.1588	1.6352	1.8688	0.5778	0.7704
5	4.0124	4.5856	1.8688	2.1024	0.963	1.1556
6	4.0124	4.5856	1.6352	1.8688	0.7704	0.963
7	3.4392	4.0124	1.6352	1.8688	0.7704	0.963
8	4.0124	4.5856	1.6352	1.8688	0.963	1.1556
9	4.5856	5.1588	1.6352	1.8688	1.3482	1.5408
10	4.5856	5.1588	1.8688	2.1024	1.3482	1.5408
11	4.0124	4.5856	1.168	1.4016	1.1556	1.3482
12	4.5856	5.1588	1.8688	2.1024	1.3482	1.5408
13	4.0124	4.5856	1.6352	1.8688	1.1556	1.3482
14	4.0124	4.5856	1.4016	1.6352	0.963	1.1556
15	3.4392	4.0124	1.6352	1.8688	0.963	1.1556
16	4.0124	4.5856	1.168	1.4016	1.3482	1.5408
17	2.2928	2.866	1.6352	1.8688	0.7704	0.963
18	3.4392	4.0124	0.9344	1.168	0.5778	0.7704
19	4.0124	4.5856	1.4016	1.6352	1.1556	1.3482
20	1.7196	2.2928	1.168	1.4016	0.5778	0.7704
21	3.4392	4.0124	1.4016	1.6352	0.7704	0.963
22	4.0124	4.5856	1.6352	1.8688	1.3482	1.5408
23	4.0124	4.5856	1.168	1.4016	1.1556	1.3482
24	2.866	3.4392	1.168	1.4016	0.5778	0.7704
25	2.866	3.4392	0.9344	1.168	0.5778	0.7704
26	1.1464	1.7196	0.7008	0.9344	0.5778	0.7704
27	1.7196	2.2928	1.168	1.4016	0.5778	0.7704
28	2.2928	2.866	1.4016	1.6352	1.1556	1.3482
29	4.0124	4.5856	1.6352	1.8688	0.7704	0.963
30	4.0124	4.5856	1.4016	1.6352	0.7704	0.963
31	4.0124	4.5856	1.4016	1.6352	0.963	1.1556

The whitenization weight functions are established for each criterion to categorize the practices into high, medium, and low importance levels. We used the following weight functions (see Figure 3.2) for this purpose.

Functions for high (f_1^1), medium (f_1^2), and low (f_1^3) categories for the criterion RME:

$$f_1^1(x) = \begin{cases} 0; & x < 0 \\ \dfrac{x}{6}; & 0 \le x \le 6 \\ 1; & x > 6 \end{cases} \tag{3.7}$$

TABLE 3.7
Final White Values

S. No.	RME	REC	RCS
1	4.916435	1.40451	1.1556
2	4.332702	1.402781	1.3482
3	2.585297	1.652582	0.7704
4	4.92223	1.644943	0.5778
5	4.338504	1.879396	0.963
6	4.341005	1.644831	0.7704
7	3.759578	1.646512	0.7704
8	4.338504	1.643182	0.963
9	4.914148	1.638692	1.3482
10	4.914148	1.874778	1.3482
11	4.335749	1.168631	1.1556
12	4.914148	1.874778	1.3482
13	4.335749	1.641349	1.1556
14	4.338504	1.406967	0.963
15	3.756197	1.644662	0.963
16	4.30019	1.168	1.34651
17	2.585297	1.652582	0.7704
18	3.762618	0.93909	0.5778
19	4.335749	1.40499	1.1556
20	1.996096	1.182738	0.5778
21	3.759578	1.409995	0.7704
22	4.332702	1.639302	1.3482
23	4.335749	1.168631	1.1556
24	3.179308	1.176916	0.5778
25	3.179308	0.940017	0.5778
26	1.3837	0.706485	0.5778
27	1.996096	1.182738	0.5778
28	2.569067	1.408319	1.1556
29	4.341005	1.644831	0.7704
30	4.341005	1.408747	0.7704
31	4.338504	1.406967	0.963

$$f_1^2(x) = \begin{cases} 0; & x\langle 0, x\rangle 6 \\ x/3; & 0 \le x \le 3 \\ (6-x)/3; & 3 < x \le 6 \end{cases} \tag{3.8}$$

$$f_1^3(x) = \begin{cases} 0; & x\langle 0, x\rangle 4 \\ 1; & 0 \le x \le 2 \\ \dfrac{4-x}{2}; & 2 < x \le 4 \end{cases} \tag{3.9}$$

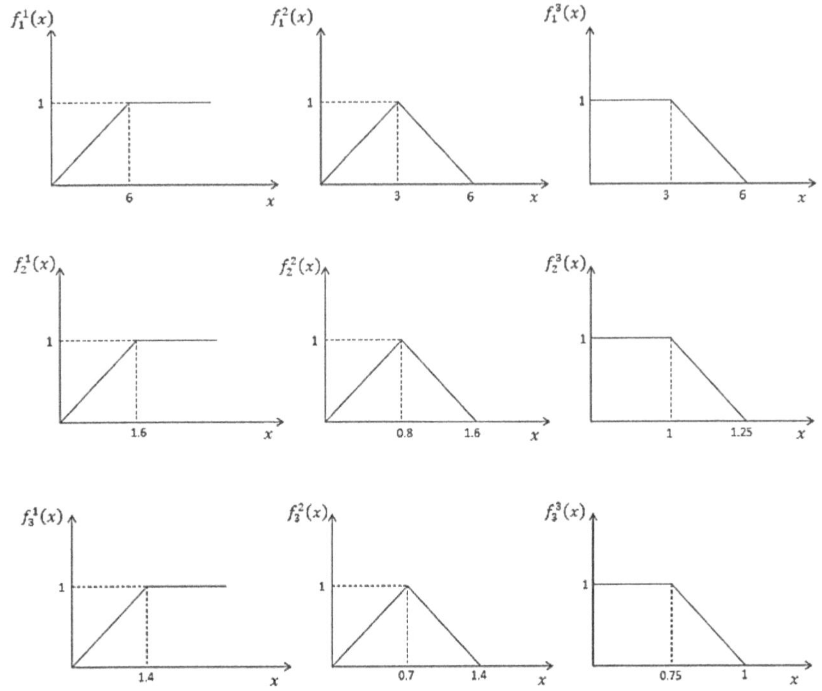

FIGURE 3.2 Whitenization weight functions.

Functions for high (f_2^1), medium (f_2^2), and low (f_2^3) categories for the criterion REC:

$$f_2^1(x) = \begin{Bmatrix} 0; & x < 0 \\ \dfrac{x}{1.6}; & 0 \le x \le 1.6 \\ 1; & x > 1.6 \end{Bmatrix} \tag{3.10}$$

$$f_2^2(x) = \begin{Bmatrix} 0; & x\langle 0, x\rangle 1.6 \\ x/0.8; & 0 \le x \le .8 \\ \dfrac{1.6 - x}{0.8}; & .8 < x \le 1.6 \end{Bmatrix} \tag{3.11}$$

$$f_2^3(x) = \begin{Bmatrix} 0; & x\langle 0, x\rangle 1.25 \\ 1.25; & 0 \le x \le 1 \\ \dfrac{(1.25 - x)}{0.25}; & 1 < x \le 1.25 \end{Bmatrix} \tag{3.12}$$

Functions for high (f_3^1), medium (f_3^2), and low (f_3^3) categories for the criterion REC:

$$f_3^1(x) = \begin{Bmatrix} 0; & x < 0 \\ \dfrac{x}{1.4}; & 0 \le x \le 1.4 \\ 1; & x > 1.4 \end{Bmatrix} \tag{3.13}$$

$$f_3^2(x) = \begin{Bmatrix} 0; & x\langle 0, x\rangle 1.4 \\ \dfrac{x}{0.7}; & 0 \le x \le 0.7 \\ \dfrac{1.4 - x}{0.7}; & .7 < x \le 1.4 \end{Bmatrix} \tag{3.14}$$

$$f_3^3(x) = \begin{Bmatrix} 0; & x\langle 0, x\rangle 1 \\ 1; & 0 \le x \le 0.75 \\ \dfrac{(1 - x)}{0.25}; & 0.75 < x \le 1 \end{Bmatrix} \tag{3.15}$$

TABLE 3.8
Cluster Coefficient Matrix

S. No.	High	Medium	Low		Category
1	0.833719	0.331361	0	0.833719	High
2	0.804197	0.390406	0	0.804197	High
3	0.586567	0.667194	0.582338	0.667194	Medium
4	0.783326	0.364903	0.1926	0.783326	High
5	0.780553	0.437694	0.028505	0.780553	High
6	0.754296	0.490209	0.176884	0.754296	High
7	0.69875	0.6013	0.245789	0.69875	High
8	0.780553	0.437694	0.028505	0.780553	High
9	0.888539	0.221723	0	0.888539	High
10	0.888539	0.221723	0	0.888539	High
11	0.743806	0.511188	0.076031	0.743806	High
12	0.888539	0.221723	0	0.888539	High
13	0.806786	0.385228	0	0.806786	High
14	0.75237	0.49406	0.028505	0.75237	High
15	0.724923	0.548953	0.098379	0.724923	High

(Continued)

TABLE 3.8 Continued

S. No.	High	Medium	Low		Category
16	0.766581	0.465638	0.076621	0.766581	High
17	0.586567	0.667194	0.582338	0.667194	Medium
18	0.576051	0.779452	0.552634	0.779452	Medium
19	0.778315	0.442171	0	0.778315	High
20	0.442862	0.662206	0.82865	0.82865	Low
21	0.671009	0.656781	0.245789	0.671009	High
22	0.832991	0.332818	0	0.832991	High
23	0.743806	0.511188	0.076031	0.743806	High
24	0.555048	0.821458	0.496099	0.821458	Medium
25	0.520461	0.890633	0.71981	0.890633	Medium
26	0.314825	0.62965	1.0578	1.0578	Low
27	0.442862	0.662206	0.82865	0.82865	Low
28	0.610024	0.614079	0.410105	0.614079	Medium
29	0.754296	0.490209	0.176884	0.754296	High
30	0.726373	0.546054	0.176884	0.726373	High
31	0.75237	0.49406	0.028505	0.75237	High

After using these weight functions, the $f_j^k(x)$ values for all the white values are obtained for $1 \le j \le 3$ and $1 \le k \le 3$. The cluster coefficients of the SCRPs are calculated using equation (3.16), and we obtained it in a matrix form of the order 31×3 containing all the cluster coefficients for all SCRPs (see Table 3.8).

$$\sigma_i^k = \Sigma_j^n f_j^k \left(\mathrm{W_{ij}}\right) * \mathrm{w_j} \tag{3.16}$$

The type of category where an SCRP has the largest cluster coefficient was identified. This particular category is selected for the SCRP in Table 3.8. Out of 31 SCRPs, 22 were in the high importance, six in the medium importance, and only three in the low importance categories. The high-importance SCRPs are extracted and used for further analysis.

3.3.3.3 TODIM

TODIM stands for *Tomada de Decisión Interactive Multicriteria* in Portuguese, roughly translating to interactive multi-criteria decision-making (Gomes and Lima, 1991). It is a multi-attribute decision-making technique that considers the subjectivity of the decision makers' behaviour for ranking alternatives using the principle of the dominance of alternatives over others. It calculates the overall prospect values of each alternative based on the global measurement of the alternatives (Tosun & Akyüz, 2015; Sen et al., 2016). Unlike other famous MCDM techniques, it doesn't consider the decision makers as rational individuals and takes care of the psychological behaviour of the decision maker while evaluating the criteria (Wu et al., 2020; D. Zhang et al., 2019). It also takes into account the acquired dominance degree of the alternatives by incorporating decision makers' attenuation factors while evaluating other options (G. Zhang et al., 2019).

The TODIM method assesses all the high-importance SCRPs based on the RME, REC, and RCS criteria. The further analysis uses the white values of the highly important practices from Table 3.7. The white values are normalized, followed by finding the relative weights of the criteria. Relative weights (w_{jr}) are calculated by dividing each criterion's weight with the maximum weight among all criteria. Then, using equations (3.17) and (3.18), the dominance of each practice over all others for the three criteria is calculated. After computing the dominance, the prospect values are calculated by adding the dominance values of SCRPs across the three criteria. Ultimately, the overall prospect values are found out by using equation (3.19). These overall prospect values are then used to rank the practices. The final ranks of all the high-importance SCRPs are shown in Table 3.9.

$$D\left(SCRP_p, SCRP_q\right) = \Sigma_{j=1}^{n} d_j\left(SCRP_p, SCRP_q\right) \tag{3.17}$$

$$d_j\left(SCRP_p, SCRP_q\right) = \begin{cases} \sqrt{\dfrac{w_{jr}\left(W_p - W_q\right)}{\Sigma_{j=1}^{n} w_{jr}}} & \text{If } SCRP_p > SCRP_q \\ 0 & \text{If } SCRP_p = SCRP_q \\ -\dfrac{1}{\theta}\sqrt{\dfrac{\Sigma_{j=1}^{n} w_{jr}\left(W_q - W_p\right)}{w_{jr}}} & \text{If } SCRP_p < SCRP_q \end{cases} \tag{3.18}$$

TABLE 3.9
Final Ranks of High-Importance SCRP

Practices	Overall Prospect Values	Ranks
SCRP 13	40.3277	1
SCRP 22	36.9712	2
SCRP 19	31.8363	3
SCRP 16	28.0934	4
SCRP 15	25.7906	5
SCRP 14	24.715	6
SCRP 10	12.9547	7
SCRP 12	12.9547	8
SCRP 9	12.54	9
SCRP 2	11.3237	10
SCRP 5	11.3115	11
SCRP 1	11.3063	12
SCRP 8	10.93	13
SCRP 6	10.2541	14
SCRP 11	10.1056	15
SCRP 4	9.97939	16
SCRP 7	9.62167	17
SCRP 21	8.89777	18
SCRP 30	7.04385	19
SCRP 23	6.70816	20
SCRP 29	6.62082	21
SCRP 31	5.04626	22

$$\xi_p = \frac{\left(\Sigma_{q=1}^{N} D\left(SCRP_p, SCRP_q\right) - \underset{p}{Min}\left(\Sigma_{q=1}^{N} D\left(SCRP_p, SCRP_q\right)\right)\right)}{\underset{p}{Max}\left(\Sigma_{q=1}^{N} D\left(SCRP_p, SCRP_q\right)\right) - \underset{p}{Min}\left(\Sigma_{q=1}^{N} D\left(SCRP_p, SCRP_q\right)\right)} \tag{3.19}$$

3.4 DISCUSSION

The use of MCDM techniques in the study is motivated by the paradox theory as it encourages the simultaneous management of the two competing strategies (SCR and SSC). Since the theory suggests avoiding the contradictory sides of the two strategies and accepting the complementary sides, the MCDM techniques help identify the conflicting and complementary practices for resilience and sustainability. Out of the three categories (high, medium, and low) obtained from grey clustering, highly important categories represent the most complementary practices. The medium- and low-importance categories represent contradictory practices. It is established that managers can start adopting practices that are ranked higher within the highly important category. Therefore, the study provided a list of practices that identify the complementary practices for resilience and sustainability. The managers can choose to adopt complementary practices and avoid contradictory practices based on the study's result. Through the adoption of appropriate practices identified in the study, they will be able to achieve SSCR. The SSCR shall enhance the quick recovery of supply chain operations at the time of disruptions like COVID-19 without compromising their sustainability goals.

Since MCDM techniques operate on the experts' opinions, there could be an influence of the experts' psychological factors on the study results. Therefore, the study proposes a human-centric decision-making technique to tackle the human psychological factors. The integrated AHP-Grey-TODIM techniques successfully manage the human-related aspects while validating the proposed decision framework. The AHP method deals with the subjectivity of experts towards sustainability criteria. The grey clustering technique controls the ambiguity that arises due to the large number and interdependency of the practices. It categorizes the practices into three categories while considering the vagueness of experts' opinions. The TODIM method incorporates the possibility of irrational behaviour by the experts during the decision-making process and ranks highly important SCRPs based on their relationship with the sustainability goals. Thus, the study proposes and validates a decision framework that manages criteria subjectivity, the ambiguity between alternatives, and the irrationality of the decision makers. Since the decision framework can control the human-related factors, the integrated framework is termed as a human-centric decision framework.

3.5 CONCLUSION

The study facilitates the operationalization of SCR in such a manner so that sustainability goals remain unaffected and the quick recovery of operations can also be ensured. This study successfully integrates SCR and SSC through MCDM techniques and results in the adoption of SSCR. The study fosters a way to establish SCR through sustainability goals, which has not been done earlier in the literature. This would lead to the better management of sustainability-related risks and empowering the SCR. The study uses an integrated MCDM technique that follows a human-centric decision-making approach. The results provide a list of SCRPs categorized and ranked under high, medium, and low importance based on their ability to achieve both resilience and sustainability goals.

The study makes several contributions to the literature as well. First, it extends the literature on integrating SCR and SSC by adopting the perspective of enhancing SCR through SSC. Second, it enriches the literature of SCR by comprehensively listing all the practices under its eight enablers. Third, it contributes to the decision-making literature by proposing a framework that considers human behavioural factors during the decision-making process. Fourth, it demonstrates the relevance of paradox theory in the supply chain literature to resolve the tensions between various competing strategies.

The study has significant implications for the managers operating in developing countries like India, where the firms are required to operate in turbulent surroundings and fulfill the sustainability goals. The study provides a comprehensive list of SCRPs. The list enables managers to operationalize SCR by choosing the right practices for their supply chain. The study also proposes a decision framework that effectively removes the SCRPs that could hinder the supply chain's sustainability goals. The managers can choose higher-ranked practices to adopt SCR without compromising sustainability goals. If managers do not intend to make major changes in their supply chains, they can start by implementing the higher-ranked practices. This would enable them to achieve SCR and SSC sequentially and gradually. This would also reduce the possibility of huge instantaneous investments.

Other than the contributions, the study also has some limitations it. The usage of a single criterion under each sustainability dimension is a critical limitation of the study. The firms may have to deal with multiple sustainability criteria under each dimension. The other limitation of the study could be that empirical research uses industries from a single country. Therefore, the generalizability of the results could be problematic. In the future, multi-objective optimization techniques can optimize the SCR and SSC, taking the paradox lens. The study can also be analyzed by using analytical modeling techniques to study the dynamics of SCR and SSC together.

REFERENCES

Abdel-Basset, M., & Mohamed, R. (2020). A novel plithogenic TOPSIS- CRITIC model for sustainable supply chain risk management. *Journal of Cleaner Production*, *247*, 119586. https://doi.org/10.1016/j.jclepro.2019.119586

Achour, N., Pantzartzis, E., Pascale, F., & Price, A. D. F. (2015). Integration of resilience and sustainability: From theory to application. *International Journal of Disaster Resilience in the Built Environment*, *6*(3), 347–362. https://doi.org/10.1108/IJDRBE-05-2013-0016

Adobor, H., & McMullen, R. S. (2018). Supply chain resilience: A dynamic and multidimensional approach. *The International Journal of Logistics Management*, *29*(4), 1451–1471. https://doi.org/10.1108/IJLM-04-2017-0093

Ageron, B., Gunasekaran, A., & Spalanzani, A. (2012). Sustainable supply management: An empirical study. *International Journal of Production Economics*, *140*(1), 168–182.

Anderies, J. M., Folke, C., Walker, B., & Ostrom, E. (2013). Aligning key concepts for global change policy: Robustness, resilience, and sustainability. *Ecology and Society*, *18*(2). https://doi.org/10.5751/ES-05178-180208

Anthony, T. (2000). Supply chain collaboration: Success in the new internet economy. *Achieving Supply Chain Excellence through Technology*, *2*, 41–44.

Autry, C. W., & Glenn Richey, R. (2009). Assessing interfirm collaboration/technology investment tradeoffs: The effects of technological readiness and organizational learning. *The International Journal of Logistics Management*, *20*(1), 30–56. https://doi.org/10.1108/09574090910954837

Aven, T. (2018). The call for a shift from risk to resilience: What does it mean? *Risk Analysis*, *0*(0). https://doi.org/10.1111/risa.13247

Barratt, M. (2004). Understanding the meaning of collaboration in the supply chain. *Supply Chain Management: An International Journal*, *9*(1), 30–42. https://doi.org/10.1108/13598540410517566

Barroso, A. P., Machado, V. H., & Machado, H. C., & V. C. (2015). Quantifying the supply chain resilience. *Applications of Contemporary Management Approaches in Supply Chains*. https://doi.org/10.5772/59580

Bellamy, L., Choudhary, A., Papaioannou, G., & Zirar, A. (2019). *Dangling between sustainability and resilience supply chain practices: Employing paradox theory to explore tensions.* https://repository.lboro.ac.uk/articles/conference_contribution/Dangling_between_sustainability_and_resilience_supply_chain_practices_employing_paradox_theory_to_explore_tensions/9499247

Blackhurst*, J., Craighead, C. W., Elkins, D., & Handfield, R. B. (2005). An empirically derived agenda of critical research issues for managing supply-chain disruptions. *International Journal of Production Research*, *43*(19), 4067–4081.

Carter, C. R., & Rogers, D. S. (2008). A framework of sustainable supply chain management: Moving toward new theory. *International Journal of Physical Distribution & Logistics Management*, *38*(5), 360–387. https://doi.org/10.1108/09600030810882816

Christopher, M., & Peck, H. (2004). Building the resilient supply chain. *The International Journal of Logistics Management*, *15*(2), 1–14. https://doi.org/10.1108/09574090410700275

Deng, X., Yang, X., Zhang, Y., Li, Y., & Lu, Z. (2019). Risk propagation mechanisms and risk management strategies for a sustainable perishable products supply chain. *Computers & Industrial Engineering*. https://doi.org/10.1016/j.cie.2019.01.014

Egels-Zandén, N., Hulthén, K., & Wulff, G. (2015). Trade-offs in supply chain transparency: The case of Nudie Jeans Co. *Journal of Cleaner Production*, *107*, 95–104. https://doi.org/10.1016/j.jclepro.2014.04.074

Elkington, J. (1997). *Cannibals with forks: The triple bottom line of 21st century business.* https://search.proquest.com/openview/804cc9d98196ef6e26d88748e89f8db0/1?cbl=35934&pq-origsite = gscholar

Eltayeb, T. K., Zailani, S., & Ramayah, T. (2011). Green supply chain initiatives among certified companies in Malaysia and environmental sustainability: Investigating the outcomes. *Resources, Conservation and Recycling*, *55*(5), 495–506. https://doi.org/10.1016/j.resconrec.2010.09.003

Fahimnia, B., & Jabbarzadeh, A. (2016). Marrying supply chain sustainability and resilience: A match made in heaven. *Transportation Research Part E: Logistics and Transportation Review*, *91*, 306–324. https://doi.org/10.1016/j.tre.2016.02.007

Fahimnia, B., Jabbarzadeh, A., & Sarkis, J. (2018). Greening versus resilience: A supply chain design perspective. *Transportation Research Part E: Logistics and Transportation Review*, *119*, 129–148. https://doi.org/10.1016/j.tre.2018.09.005

FICCI. (2020). *Impact of coronavirus on Indian businesses.* FICCI. https://ficci.in/SEDocument/20497/FICCI-Survey-COVID19.pdf

Fiksel, J. (2006). Sustainability and resilience: Toward a systems approach. *Sustainability: Science, Practice and Policy*, *2*(2), 14–21. https://doi.org/10.1080/15487733.2006.11907980

Ghadge, A., Wurtmann, H., & Seuring, S. (2020). Managing climate change risks in global supply chains: A review and research agenda. *International Journal of Production Research*, *58*(1), 44–64. https://doi.org/10.1080/00207543.2019.1629670

Giannakis, M., & Papadopoulos, T. (2016). Supply chain sustainability: A risk management approach. *International Journal of Production Economics*, *171*, 455–470. https://doi.org/10.1016/j.ijpe.2015.06.032

Gomes, L. F. A. M., & Lima, M. M. P. P. (1991). TODIMI: Basics and application to multi-criteria ranking. *Foundations of Computing and Decision Sciences*, *16*(3–4), 113-127.

Govindan, K., Khodaverdi, R., & Jafarian, A. (2013). A fuzzy multi criteria approach for measuring sustainability performance of a supplier based on triple bottom line approach. *Journal of Cleaner Production*, *47*, 345–354. https://doi.org/10.1016/j.jclepro.2012.04.014

Handmer, J. W., & Dovers, S. R. (1996). A typology of resilience: Rethinking institutions for sustainable development. *Industrial & Environmental Crisis Quarterly*, *9*(4), 482–511. https://doi.org/10.1177/108602669600900403

Harmon, R. R., & Cowan, K. R. (2009). A multiple perspectives view of the market case for green energy. *Technological Forecasting and Social Change*, *76*(1), 204–213. https://doi.org/10.1016/j.techfore.2008.03.026

He, L., Wu, Z., Xiang, W., Goh, M., Xu, Z., Song, W., Ming, X., & Wu, X. (2020). A novel Kano-QFD-DEMATEL approach to optimise the risk resilience solution for sustainable supply chain. *International Journal of Production Research*, *0*(0), 1–22. https://doi.org/10.1080/00207543.2020.1724343

Hofmann, H., Busse, C., Bode, C., & Henke, M. (2014). Sustainability-related supply chain risks: Conceptualization and management. *Business Strategy and the Environment*, *23*(3), 160–172. https://doi.org/10.1002/bse.1778

Hosseini-Motlagh, S.-M., Samani, M. R. G., & Shahbazbegian, V. (2020). Innovative strategy to design a mixed resilient-sustainable electricity supply chain network under uncertainty. *Applied Energy*, *280*, 115921. https://doi.org/10.1016/j.apenergy.2020.115921

Irvin, S. (2020). *Irvine, Samuel.pdf* [University of Wisconsin]. https://minds.wisconsin.edu/bitstream/handle/1793/80249/Irvine,%20Samuel.pdf?sequence=1

Jabbarzadeh, A., Fahimnia, B., & Sabouhi, F. (2018). Resilient and sustainable supply chain design: Sustainability analysis under disruption risks. *International Journal of Production Research*, *0*(0), 1–24. https://doi.org/10.1080/00207543.2018.1461950

John, E. (1997). *Cannibals with forks: The triple bottom line of 21st century business.* Stony Creek, CT: New Society.

Karutz, R., Riedner, L., Vega, L. R., Stumpf, L., & Damert, M. (2018). Compromise or complement? Exploring the interactions between sustainable and resilient supply chain management. *International Journal of Supply Chain and Operations Resilience*, *3*(2), 117. https://doi.org/10.1504/IJSCOR.2018.090768

Kaur, H., Singh, S. P., Garza-Reyes, J. A., & Mishra, N. (2020). Sustainable stochastic production and procurement problem for resilient supply chain. *Computers & Industrial Engineering*, *139*, 105560. https://doi.org/10.1016/j.cie.2018.12.007

Kayikci, Y. (2020). Stream processing data decision model for higher environmental performance and resilience in sustainable logistics infrastructure. *Journal of Enterprise Information Management*, *ahead-of-print*(ahead-of-print). https://doi.org/10.1108/JEIM-08-2019-0232

Klassen, R. D., & Vereecke, A. (2012). Social issues in supply chains: Capabilities link responsibility, risk (opportunity), and performance. *International Journal of Production Economics*, *140*(1), 103–115. https://doi.org/10.1016/j.ijpe.2012.01.021

Kovács, G., & Halldórsson, Á. (2010). The sustainable agenda and energy efficiency: Logistics solutions and supply chains in times of climate change. *International Journal of Physical Distribution & Logistics Management*, *40*(1/2), 5–13. https://doi.org/10.1108/09600031011018019

Li, X., Goldsby, T. J., & Holsapple, C. W. (2009). Supply chain agility: Scale development. *The International Journal of Logistics Management*, *20*(3), 408–424. https://doi.org/10.1108/09574090911002841

Lin, C.-H., Wu, C.-H., & Huang, P.-Z. (2009). Grey clustering analysis for incipient fault diagnosis in oil-immersed transformers. *Expert Systems with Applications*, *36*(2), 1371–1379.

Lummus, R. R., Vokurka, R. J., & Duclos, L. K. (2005). Delphi study on supply chain flexibility. *International Journal of Production Research*, *43*(13), 2687–2708. https://doi.org/10.1080/00207540500056102

Manuj, I., & Mentzer, J. T. (2008). Global supply chain risk management strategies. *International Journal of Physical Distribution & Logistics Management*, *38*(3), 192–223.

Marchese, D., Reynolds, E., Bates, M. E., Morgan, H., Clark, S. S., & Linkov, I. (2018). Resilience and sustainability: Similarities and differences in environmental management applications. *Science of The Total Environment*, *613–614*, 1275–1283. https://doi.org/10.1016/j.scitotenv.2017.09.086

Mohammed, A. (2020). Towards 'gresilient' supply chain management: A quantitative study. *Resources, Conservation and Recycling*, *155*, 104641. https://doi.org/10.1016/j.resconrec.2019.104641

Mohammed, A., Harris, I., Soroka, A., & Nujoom, R. (2019). A hybrid MCDM-fuzzy multi-objective programming approach for a G-resilient supply chain network design. *Computers & Industrial Engineering*, *127*, 297–312. https://doi.org/10.1016/j.cie.2018.09.052

Mousavi Ahranjani, P., Ghaderi, S. F., Azadeh, A., & Babazadeh, R. (2020). Robust design of a sustainable and resilient bioethanol supply chain under operational and disruption risks. *Clean Technologies and Environmental Policy*, *22*(1), 119–151. https://doi.org/10.1007/s10098-019-01773-2

Ponomarov, S. Y., & Holcomb, M. C. (2009). Understanding the concept of supply chain resilience. *The International Journal of Logistics Management*, *20*(1), 124–143. https://doi.org/10.1108/09574090910954873

Pun, K. F., & Hui, I. K. (2001). An analytical hierarchy process assessment of the ISO 14001 environmental management system. *Integrated Manufacturing Systems*, *12*(5), 333–345. https://doi.org/10.1108/EUM0000000005711

Rajesh, R. (2018a). On sustainability, resilience, and the sustainable – resilient supply networks. *Sustainable Production and Consumption*, *15*, 74–88. https://doi.org/10.1016/j.spc.2018.05.005

Rajesh, R. (2018b). Measuring the barriers to resilience in manufacturing supply chains using Grey Clustering and VIKOR approaches. *Measurement*, *126*, 259–273. https://doi.org/10.1016/j.measurement.2018.05.043

Ramezankhani, M. J., Torabi, S. A., & Vahidi, F. (2018). Supply chain performance measurement and evaluation: A mixed sustainability and resilience approach. *Computers & Industrial Engineering*, *126*, 531–548. https://doi.org/10.1016/j.cie.2018.09.054

Reeves, M., Lang, N., & Carlsson-Szlezak, P. (2020). Lead Your Business Through the Coronavirus Crisis. *Harvard Business Review*, 7.

Rice, J. B. Jr, & Sheffi, Y. (2005). A Supply Chain View of the Resilient Enterprise. *MIT Sloan Management Review; Cambridge*, *47*(1), 41–48.

Roehrich, J. K., Grosvold, J., & Hoejmose, S. (2014). Reputational risks and sustainable supply chain management: Decision making under bounded rationality. *International Journal of Operations & Production Management*, *34*(5), 695–719. https://doi.org/10.1108/IJOPM-10-2012-0449

Saaty, R. W. (1987). The analytic hierarchy process – What it is and how it is used. *Mathematical Modelling*, *9*(3–5), 161–176.

Saaty, T. L. (2008). Decision making with the analytic hierarchy process. *International Journal of Services Sciences*, *1*(1), 83–98.

Scholten, K., & Schilder, S. (2015). The role of collaboration in supply chain resilience. *Supply Chain Management: An International Journal*, *20*(4), 471–484. https://doi.org/10.1108/SCM-11-2014-0386

Sen, D. K., Datta, S., & Mahapatra, S. S. (2016). Application of TODIM (Tomada de Decisión IneRativa Multicritero) for industrial robot selection. *Benchmarking: An International Journal*, *23*(7), 1818–1833. https://doi.org/10.1108/BIJ-07-2015-0078

Seuring, S., & Müller, M. (2008). From a literature review to a conceptual framework for sustainable supply chain management. *Journal of Cleaner Production*, *16*(15), 1699–1710. https://doi.org/10.1016/j.jclepro.2008.04.020

Shad, M. K., Lai, F.-W., Fatt, C. L., Klemeš, J. J., & Bokhari, A. (2019). Integrating sustainability reporting into enterprise risk management and its relationship with business performance: A conceptual framework. *Journal of Cleaner Production*, *208*, 415–425. https://doi.org/10.1016/j.jclepro.2018.10.120

Shamout, S., Boarin, P., & Wilkinson, S. (2021). The shift from sustainability to resilience as a driver for policy change: A policy analysis for more resilient and sustainable cities in Jordan. *Sustainable Production and Consumption*, *25*, 285–298. https://doi.org/10.1016/j.spc.2020.08.015

Sheffi, Y., Rice, J. B., Fleck, J. M., & Caniato, F. (2003). *Supply chain response to global terrorism: A situation scan.* Center for Transportation and Logistics, MIT, Department of Management, Economics and Industrial Engineering, Politecnico Di Milano, EurOMA POMS Joint International Conference.

Slaper, T. F., & Hall, T. J. (2011). The triple bottom line: What is it and how does it work. *Indiana Business Review*, *86*(1), 4–8.

Smith, W. K. (2014). Dynamic Decision Making: A Model of Senior Leaders Managing Strategic Paradoxes. *Academy of Management Journal*, *57*(6), 1592–1623. https://doi.org/10.5465/amj.2011.0932

Smith, W. K., & Lewis, M. W. (2011). Toward a Theory of Paradox: A Dynamic equilibrium Model of Organizing. *Academy of Management Review*, *36*(2), 381–403. https://doi.org/10.5465/amr.2009.0223

Smith, W. K., & Tracey, P. (2016). Institutional complexity and paradox theory: Complementarities of competing demands. *Strategic Organization*, *14*(4), 455–466. https://doi.org/10.1177/1476127016638565

Suresh, N. C., & Kaparthi, S. (1992). Flexible automation investments: A synthesis of two multi-objective modeling approaches. *Computers & Industrial Engineering*, *22*(3), 257–272. https://doi.org/10.1016/0360-8352(92)90004-4

Tarei, P. K., Thakkar, J. J., & Nag, B. (2018). A hybrid approach for quantifying supply chain risk and prioritizing the risk drivers: A case of Indian petroleum supply chain. *Journal of Manufacturing Technology Management*, *29*(3), 533–569. https://doi.org/10.1108/JMTM-10-2017-0218

The Sustainability Consortium. (2016). *Greening global supply chains: From blind spots to hotspots to action. Impact Report.* https://www.Sustainabilityconsortium. org/wpcontent/themes/enfold-child/assets/pdf/2016-impact-report.pdf.

Thuermer, K. E. (2008). Air cargo braces for a slowdown. *Logistics Management*, *47*(11).

Tosun, Ö., & Akyüz, G. (2015). A fuzzy TODIM approach for the supplier selection problem. *International Journal of Computational Intelligence Systems*, *8*(2), 317–329. https://doi.org/10.1080/18756891.2015.1001954

Tsang, S., Welford, R., & Brown, M. (2009). Reporting on community investment. *Corporate Social Responsibility and Environmental Management*. https://onlinelibrary.wiley.com/doi/pdf/10.1002/csr.178

Valinejad, F., & Rahmani, D. (2018). Sustainability risk management in the supply chain of telecommunication companies: A case study. *Journal of Cleaner Production*, *203*, 53–67. https://doi.org/10.1016/j.jclepro.2018.08.174

Walker, D. H., Bourne, L., & Rowlinson, S. (2007). Stakeholders and the supply chain. In *Procurement systems: A Cross Industry Project Management Perspective* (pp. 70–100). Routledge.

WCED. (1987). *Our common future: The world commission on environment*. http://www.un-documents.net/our-common-future.pdf

Wen, K.-L. (2008). A Matlab toolbox for grey clustering and fuzzy comprehensive evaluation. *Advances in Engineering Software*, *39*(2), 137–145. https://doi.org/10.1016/j.advengsoft.2006.12.002

Wieland, A., & Marcus Wallenburg, C. (2013). The influence of relational competencies on supply chain resilience: A relational view. *International Journal of Physical Distribution & Logistics Management*, *43*(4), 300–320.

Wijethilake, C., & Lama, T. (2019). Sustainability core values and sustainability risk management: Moderating effects of top management commitment and stakeholder pressure. *Business Strategy and the Environment*, *28*(1), 143–154. https://doi.org/10.1002/bse.2245

Wind, Y., & Saaty, T. L. (1980). Marketing applications of the analytic hierarchy process. *Management Science*, *26*(7), 641–658. https://doi.org/10.1287/mnsc.26.7.641

Wu, Y., Wu, C., Zhou, J., Zhang, B., Xu, C., Yan, Y., & Liu, F. (2020). A DEMATEL-TODIM based decision framework for PV power generation project in expressway service area under an intuitionistic fuzzy environment. *Journal of Cleaner Production*, *247*, 119099. https://doi.org/10.1016/j.jclepro.2019.119099

Zahiri, B., Zhuang, J., & Mohammadi, M. (2017). Toward an integrated sustainable-resilient supply chain: A pharmaceutical case study. *Transportation Research Part E: Logistics and Transportation Review*, *103*, 109–142. https://doi.org/10.1016/j.tre.2017.04.009

Zailani, S., Jeyaraman, K., Vengadasan, G., & Premkumar, R. (2012). Sustainable supply chain management (SSCM) in Malaysia: A survey. *International Journal of Production Economics*, *140*(1), 330–340. https://doi.org/10.1016/j.ijpe.2012.02.008

Zamanian, M. R., Sadeh, E., Sabegh, Z. A., & Rasi, R. E. (2020). A multi-objective optimization model for the resilience and sustainable supply chain: A case study. *International Journal of Supply and Operations Management; Tehran*, *7*(1), 51–75.

Zhang, D., Li, Y., & Wu, C. (2019). An extended TODIM method to rank products with online reviews under intuitionistic fuzzy environment. *Journal of the Operational Research Society*, *0*(0), 1–13. https://doi.org/10.1080/01605682.2018.1545519

Zhang, G., Wang, J., & Wang, T. (2019). Multi-criteria group decision-making method based on TODIM with probabilistic interval-valued hesitant fuzzy information. *Expert Systems*, e12424. https://doi.org/10.1111/exsy.12424

Zhou, Z., Cheng, S., & Hua, B. (2000). Supply chain optimization of continuous process industries with sustainability considerations. *Computers & Chemical Engineering*, *24*(2–7), 1151–1158. https://doi.org/10.1016/S0098-1354(00)00496-8

4 Addressing the Strategies for the Sustainable Supply Chain in Post-COVID-19 Pandemic

Subhodeep Mukherjee, Manish Mohan Baral, Venkataiah Chittipaka, and Surya Kant Pal

CONTENTS

4.1 INTRODUCTION

Supply chains (SC) are the foundation of economies and societies and are linked to nature. Shared interrelations and inputs between SCs set off the connections in these SC systems, which are unpredictable. In either case, SCs' obligations, versatility, and supportability have to be scrutinized in 2021. The COVID-19 virus (CV-19) epidemic and worldwide pandemic, another provocateur of SC disruptions, has resulted in an unusual arrangement of shocks for SCs worldwide (Al-Zabidi et al., 2021; Queiroz

DOI: 10.1201/9781003150084-4

et al., 2020). The CV-19 epidemic and the global pandemic significantly impacted all economic and social zones, resulting in a series of entirely new complex settings for SC analysts and professionals.

Recently, a novel, long-running troublesome pandemic known as CV-19 had a significant impact on global SC (Işık et al., 2021; Singh et al., 2020). Currently, the CV-19 pandemic is a health emergency, as well as an emergency in the labor market and financial sectors (Bochtis et al., 2020; M. Sharma et al., 2021). The CV-19 pandemic has wreaked havoc on the SC all over the world. SC disruptions are increasing rapidly, thanks to public authority-ordered lockdowns and tight movement restrictions (Barman et al., 2021; Remko, 2020; Shanmugam et al., 2020).

Many Fortune 1000 companies have experienced SC disruptions caused by CV-19 (Biswas & Das, 2020; Richards & Rickard, 2020; Sarkis et al., 2020). Chinese fares fell by about 17% at the peak of the pandemic, and this figure is expected to rise to 32% by 2020. According to supply chain network Tradeshift, standard global installment terms increased by 1.7% to 37.4 days in the first quarter of 2020, up from 36.7 days in 2019. Epidemic outbreaks are classified as uncommon SC hazards because they occur infrequently and have long-term effects on SCs (M. Sharma et al., 2021; R. Sharma et al., 2020). As a result, predicting the impact of CV-19 on SC is extremely difficult for specialists and experts. The research questions addressed in this study are:

RQ1: What strategies are required for the sustainable supply chain (SSC) pre- and post-CV-19 pandemic?
RQ2: How can these help industries emerging from the pandemic?

This study identifies the strategies to overcome the CV-19 pandemic. Six strategies are recognized for the study, mainly disruption risk management capacity (DRMC), supply chain agility (SCA), delivery reliability (DR), strong legislation facility to tackle CV-19 (LF), customer support awareness (CSA), and supply chain partner collaboration (SCPC). These six strategies are treated as independent variables, and one dependent variable, sustainable supply chain, is used.

4.2 LITERATURE REVIEW

The outbreak of CV-19 has disrupted most SC facilities worldwide (Buhalis & Leung, 2018; Karmaker et al., 2021; Remko, 2020). SC interruptions occur because of occasions that have a low likelihood of happening; however, there is an exceptionally high effect, such as that due to catastrophic events (floods, earthquakes, and so on), psychological oppressor attacks, and pandemics (SARS, Ebola, swine influenza, CV-19, and so on) (Hobbs, 2020; Xu et al., 2020). Unlike operational dangers that include an average postponement in the stockpile, machine failure, or changes in demand, disturbance hazards have gradually expanded their influence as the outcomes permeate through the whole SC and influence the business activities of the human populace (Karmaker et al., 2021; Trotter et al., 2020). Supportability, generally acknowledged as "advancement that addresses the present issues without trading off the capacity of people in the future to address their issues," is a unique drawn-out cycle and represents a few difficulties for SC managers (Raut et al., 2017). Along these lines, a SSC tries to decrease SC tasks' adverse effects and improve associations' social, financial, and ecological execution (Schniederjans et al., 2020). Business associations have likewise started to incorporate SSC as an approach to improve their image. Also, SSC mitigates dangers and weaknesses, such as ecological harm and work deficiencies, which can support business strength and lessen postponements and expenses in creating and disseminating measures (A. Sharma et al., 2020). In any case, various organizations in arising economies today don't have the aptitude needed for fruitful usage and transformation of supportability; this is fundamental because manageability research is not universally characterized, created, or applied (Mangla et al., 2018). The aforementioned remarks mirror the need to talk about maintainability in SCs to set up reasonableness and give possibilities to economic execution in creating economies (Savary et al., 2020).

For specific SCs (e.g., face covers, hand sanitizer, cleaning shower), the request has expanded, and supply couldn't adapt to that circumstance (Fink, 2020). For different SCs (e.g., auto industry), requests and supplies have radically dropped, resulting in the peril of liquidations and the need for legislative backing (Metwally et al., 2020). Here, the inquiries of SC survivability again emerge (A. Sharma et al., 2020). It is evident that both of these inquiries go beyond the current cutting edge in SC risk, maintainability, and flexibility since this can't be settled separately inside every one of these points of view. Coordinated systems and augmentation are required when long haul, severe worldwide disturbances influence all components of SC environments (i.e., organizations, society, nature, and economies) (Chowdhury et al., 2020; Gray, 2020).

Developing vulnerabilities and difficulties connected with SSC have forced uncommon accentuation on their economies, considering difficult economic situations in various agricultural nations (Kumar et al., 2021; Gössling et al., 2021). Improper checking of these moves prompts severe aggravations in SC organizations and subsequently social orders. These unsettling influences can cause perpetual financial harm. For example, the CV-19 pandemic has produced a tremendous interruption in SC globally (Kalogiannidis & Melfou, 2020; Metwally et al., 2020; Qingbin et al., 2020). Such aggravations negatively impact deals income, profitability, acquirement procedures, brand picture, materials supply, partner and client well-being, coordination administration, and generally speaking, execution in SC. These negative impacts are brought about by the quick outcomes of reach SC segments, including acquisition, creation, circulation, and coordination (Ivanov, 2020a; Ivanov & Das, 2020). Likewise, the administration of SC measures has been amazingly troublesome, as specific periods of SC exercises have been halted.

4.3 RESEARCH HYPOTHESIS AND STRATEGIES FOR SSC POST-CV-19

SSC drivers guarantee productive activity just as practiced by monetary execution, risk management, speedy reactions to uncertain conditions, the satisfaction of the maintainability assumptions, and manageability (Sajjad et al., 2020). There have been secondary considerations on key supportability drivers for SSC effective execution.

It is essential to analyze the drivers of supportability, particularly in handling the pandemic's impacts on SC. Such drivers help organizations upgrade manageability activities and generally improve maintainability execution.

The strategies for SSC post-CV-19 pandemic are:

3.1. *Disruption Risk Management Capacity (DRMC)*: Interruption hazard the executives limit empowers the organizations to seek after the way of life towards the production of persistent danger appraisal groups because of the drawn-out impact of CV-19 on the store network (i.e., the firms need to build a capacity like in inventory to handle any kind of disruptions in nature) (Amalia et al., 2020; Chowdhury et al., 2020).

H1: DRMC positively impacts the SSC.

3.2. *Supply Chain Agility (SCA)*: Agility in SC builds the network's visibilities inside the creation and the dissemination organizations (i.e., supply chain network should be transparent) to maintain stock to fluctuating business sector interest (i.e., market share of the company) during pandemics (Gray, 2020; Shahed et al., 2021).

H2: SCA positively impacts the SSC.

3.3. *Delivery Reliability (DR)*: DR during CV-19 fulfills the clients' necessities and influences the production network supportability (Barman et al., 2021; Nakat & Bou-Mitri, 2021).

H3: DR positively impacts the SSC.

3.4. *Strong Legislation Facility to Tackle CV-19 (LF)*: Regulation to tie the associations to embrace manageability concerning work relations, business conditions, and natural administration during CV-19 (Barman et al., 2021; M. Sharma et al., 2021).

H4: LF positively impacts the SSC.

3.5. *Customer Support Awareness (CSA)*: Customers' consciousness of supportable items has pressed associations to embrace availability (Roggeveen & Sethuraman, 2020; Sajjad et al., 2020; Sarkis et al., 2020).

H5: CSA positively impacts the SSC.

3.6. *Supply Chain Partners Collaboration (SCPC)*: Collaborative planning among the SC partners guarantees smooth material and creation stream (Adivar et al., 2019; Cohen, 2020; Karmaker et al., 2021).

H6: SCPC positively impacts the SSC.

4.4 RESEARCH METHODOLOGY

4.4.1 Sampling

The sampling method used is simple random sampling as it gives chances to address all the samples taken for the study. A questionnaire was developed with the help of academicians and researchers for survey-based research. A seven-point Likert scale was used for preparing the structured questionnaire (Baral & Verma, 2021; Mukherjee & Chittipaka, 2021). An online survey was carried out due to the pandemic, which saved time by not having to visit the industries (Baral et al., 2021; Pal et al., 2021). The target population was the employees working in five industrial sectors in India – automobile, electronics, retail, textile, and chemical. The total number of questionnaires sent was 656, but only 396 questionnaires were filled out appropriately and were valid for the analysis. Software used for the analysis was SPSS 20.0 and AMOS 22.0.

4.4.2 Demographics of the Respondents

Table 4.1 shows the demographics of the respondents. The survey was conducted in five various industries. Twenty-four percent of respondents were from the textile industries, 22% from the automobile sector, 21% from the electronics industry, 18% from the chemical industry, and 15% from the retail sector. Thirty-eight percent of respondents were plant managers, 27% were purchase managers, 21% were supply chain managers, and 14% were store managers.

4.5 DATA ANALYSIS AND RESULT

4.5.1 Common Method Bias

Harman's single-factor test was used. Things (Items) were put through an unrotated exploratory factor review to see if a single factor emerged or accounted for most of

TABLE 4.1
Demographics of Respondents

Characteristics	Percentage
Gender	
Male	43%
Female	57%
Industry	
Automobile	22%
Electronics	21%
Retail	15%
Textile	24%
Chemical	18%
Respondents' current position	
Plant Manager	38%
Supply Chain Manager	21%
Store Manager	14%
Purchase Manager	27%

the variance. The first variable with the highest eigenvalue explains 19.522% of our test variance, less than half of the total variance (Podsakoff et al., 2003).

4.5.2 Cronbach's Alpha

The reliability and validity tests were carried out for the six factors taken for the study. Cronbach's alpha value was calculated, and the values should be greater than 0.70, which is the recommended threshold level (Nunnally, 1994).

4.5.3 Composite Reliability (CR)

CR is calculated for all six factors. CR measures the internal consistency of the scale. The threshold level should be greater than 0.70, the accepted value for the factors (Henseler et al., 2009).

4.5.4 Exploratory Factor Analysis (EFA)

EFA is measured using the software SPSS 20.0. The value of KMO (Kaiser-Meyer-Olkin) test is 0.760. The principal axis factoring extraction method was used. Only eigenvalues more significant than 1 are extracted because it explains the most variance. The percent variance for the first component is 22.788%, for the second component is 14.043%, for the third component is 11.404%, for the fourth component is 8.653%, for the fifth component is 8.381%, and for the sixth component is 6.570%. The overall percentage for all six factors is 71.839%.

In EFA, a rotated component matrix is calculated to group the indicators in meaningful factors. The method of extraction followed is the principal component, and the rotation method selected is varimax rotation. The values of the rotated component matrix are shown in Table 4.2.

TABLE 4.2
Cronbach's Alpha, Composite Reliability, Rotated Component Matrix, Average Variance Extracted

Latent Variable	Indicators	Rotated Component Matrix	Cronbach's Alpha (α)	Composite Reliability (CR)	Average Variance Extracted (AVE)
LF	LF1	.844	0.882	0.921	0.891
	LF2	.929			
	LF3	.902			
CSA	CSA1	.836	0.839	0.901	0.865
	CSA2	.950			
	CSA3	.811			
SCPC	SCPC1	.770	0.778	0.859	0.818
	SCPC2	.874			
	SCPC3	.809			

Latent Variable	Indicators	Rotated Component Matrix	Cronbach's Alpha (α)	Composite Reliability (CR)	Average Variance Extracted (AVE)
DR	DR1	.731	0.732	0.845	0.802
	DR2	.855			
	DR3	.820			
DRMC	DRMC1	.840	0.73	0.845	0.803
	DRMC2	.804			
	DRMC3	.766			
SCA	SCA1	.825	0.722	0.841	0.796
	SCA2	.866			
	SCA3	.697			

4.5.5 Confirmatory Factor Analysis (CFA)

Figure 4.1 represents the CFA. The independent variables used in the model are LF: Strong Legislation Facility to Tackle CV-19 has three indicators – LF1, LF2, and LF3; CSA: Customer Support Awareness has three indicators – CSA1, CSA2, and CSA3; SCPC: Supplier Side has three indicators – SCPC1, SCPC2, and SCPC3; DR: Delivery Reliability has three indicators – DR1, DR2, and DR3; DRMC: Disruption Risk Management Capacity has three indicators – DRMC1, DRMC2, and DRMC3; and SCA: Supply Chain Agility has three indicators – SCA1, SCA2, and SCA3. Table 4.3 shows the parameters for the model, and all the parameters satisfied the threshold level.

4.5.6 Construct Validity (CV)

An essential logical idea is to evaluate the validity of a CV development measure. CV is the extent to which a test quantifies the concept of development that it is supposed to quantify. CV is usually tested by assessing the relationship on a few scales. CV does not feature a cut-off (DeVellis et al., 2003).

4.5.7 Divergent or Discriminant Validity

The construct correlation matrix for measuring the discriminant validity is shown in Table 4.4. The extracted variance and its squared correlation for the LF is 0.772 for LF and CSA, 0.255 for the variance; LF and SCPC, 0.732 and 0.192; LF and DR, 0.719 and 0.0003; LF and DRMC, 0.720 and 0.001; and LF and SCA, 0.714 and 0.0005. For the CSA it is 0.709 and 0.009 for CSA and SCPC; 0.696 and 0.005 for CSA and DR; 0.697 for CSA and DRMC; 0.696 and 0.000 for DR and CSA; and 0.691 and 0.0003 for SCA, SCPC, and DR. As such, the validity is divergent or discriminant.

The load estimate is above 0.50, and standard errors are below +2.5 with the limit level. The load estimate is below 0.50 (Hair et al., 2013). The critical ratios of Hair et al. are more than 1.96 (2013). This is less than the threshold of 0.05, and P-values are 0.000. Statistically critical are the loads. Therefore, the requirements are met, and the final model can be built. The path analysis result for the CFA is shown in Table 4.5.

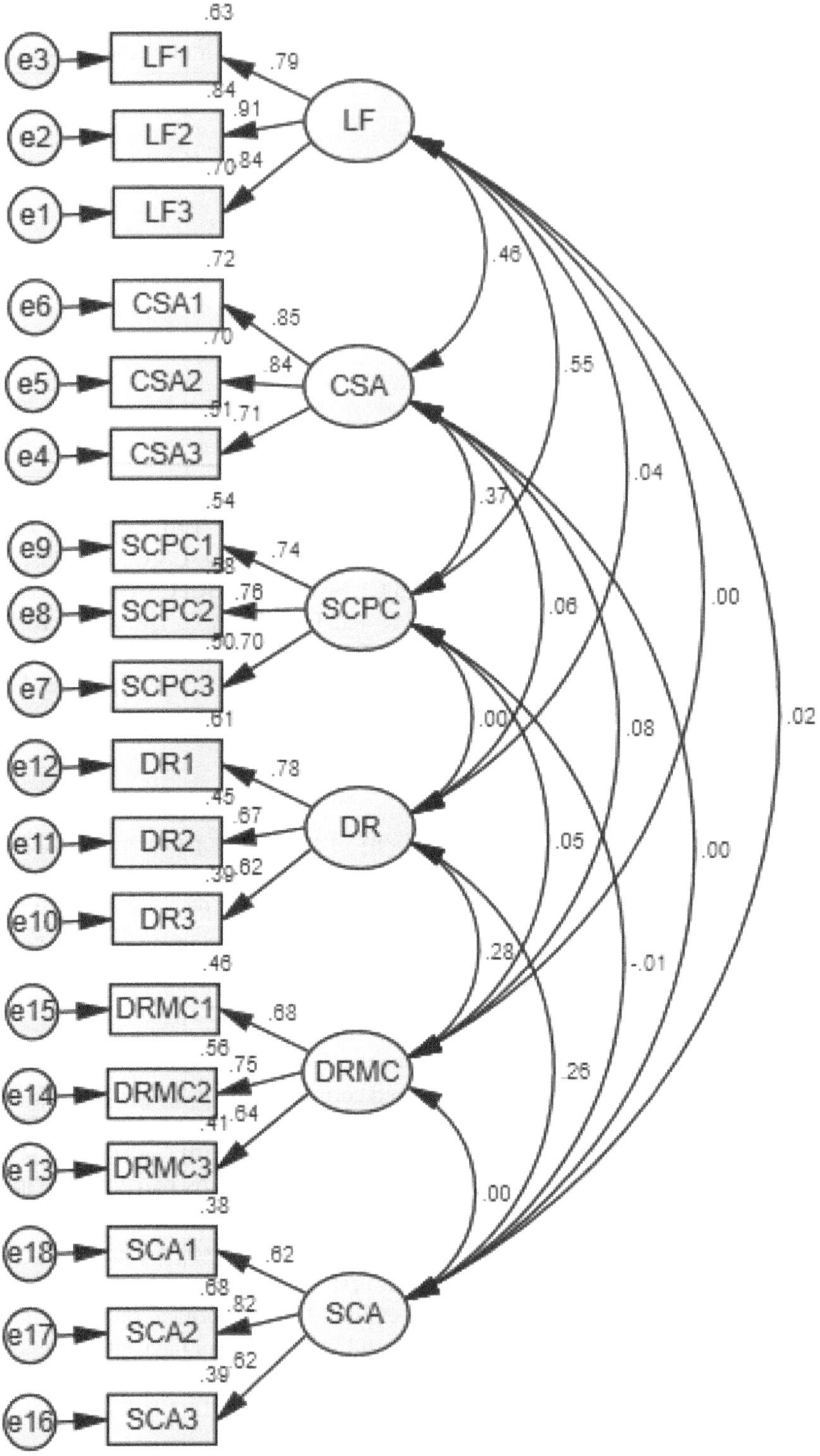

FIGURE 4.1 Confirmatory factor analysis for the latent variables.

TABLE 4.3
Model Fit Measures for the Confirmatory Factor Analysis

Goodness-of-Fit Indices	Default Model	Benchmark
Absolute goodness-of-fit measure		
χ2/df (CMIN/DF)	2.321	Lower Limit:1.0 Upper Limit 2.0/3.0 or 5.0
GFI	0.929	>0.90
Incremental fit measure		
CFI	0.938	⩾0.90
IFI	0.939	⩾0.90
TLI	0.921	⩾0.90
Parsimony fit measure		
PCFI	0.736	⩾0.50
PNFI	0.704	⩾0.50

TABLE 4.4
Construct Correlation Matrix

	Variance Extracted Between Factors					
	LF	CSA	SCPC	DR	DRMC	SCA
LF	1					
CSA	0.772	1				
SCPC	0.732	0.709	1			
DR	0.719	0.696	0.656	1		
DRMC	0.720	0.697	0.657	0.644	1	
SCA	0.714	0.691	0.651	0.639	0.64	1

TABLE 4.5
Path Analysis Result for Confirmatory Factor Analysis

	Estimate	SE	CR	P
LF3<---LF	0.839			
LF2<---LF	0.915	0.054	16.94	0.000
LF1<---LF	0.794	0.058	13.69	0.000
CSA3<---CSA	0.712			
CSA2<---CSA	0.837	0.083	10.08	0.000
CSA1<---CSA	0.847	0.080	10.59	0.000
SCPC3<---SCPC	0.704			
SCPC2<---SCPC	0.763	0.094	8.12	
SCPC1<---SCPC	0.738	0.089	8.29	0.000
DR3<---DR	0.623			
DR2<---DR	0.671	0.106	6.33	0.000
DR1<---DR	0.782	0.161	4.86	0.000

(Continued)

TABLE 4.5 Continued

	Estimate	SE	CR	P
DRMC3<---DRMC	0.643			
DRMC2<---DRMC	0.749	0.128	5.85	0.000
DRMC1<---DRMC	0.678	0.129	5.26	0.000
SCA3<---SCA	0.622			
SCA2<---SCA	0.823	0.145	5.67	0.000
SCA1<---SCA	0.617	0.103	5.99	0.000

4.5.8 Structural Equation Modeling (SEM)

For testing all the proposed six hypotheses, we have used the SEM approach (Byrne, 2010). Figure 4.2 represents the final model. Independent variables and dependent variables along with its indicators are LF: Strong Legislation Facility to Tackle CV-19 has three indicators – LF1, LF2, and LF3; CSA: Customer Support Awareness has three indicators – CSA1, CSA2, and CSA3; SCPC: Supplier Side has three indicators – SCPC1, SCPC2, and SCPC3; DR: Delivery Reliability has three indicators – DR1, DR2, and DR3; DRMC: Disruption Risk Management Capacity has three indicators – DRMC1, DRMC2, and DRMC3; and SCA: Supply Chain Agility has three indicators – SCA1, SCA2, and SCA3. One dependent variable is SSC: Sustainable Supply Chain, which has four indicators – SSC1, SSC2, SSC3, and SSC4.

Table 4.7 shows the path estimate analysis results. The result shows the six P-value hypotheses (Hair et al. 2010). Therefore, SSC positively impacts LF, CSA, SCPC, DR, DRMC, and SCA. The multiple square correlations (R^2) help measure how well a regressive line estimates the actual data points between zero and zero (Hair et al., 2010). The closer to 1 it is, the more technological predictability for the model (Kline, 2015). Therefore, 65% of the SSC variance can be explained in the proposed model.

4.6 DISCUSSION

The CV-19 pandemic has uncovered many areas of weakness in the sustainability of several emerging economies' SCs (Cariappa et al., 2020; Rowan & Galanakis, 2020; Sarkis, 2020). To formulate and enforce these policies effectively, businesses would need government intervention. The pandemic stressed government subsidy's value in joint research projects involving retailers and manufacturers to make the supply chain more competitive. During the CV-19 pandemic, however, government assistance took the form of incentives, tax reductions, and lesser interest loans for business activity (Annosi et al., 2021; Barman et al., 2021; Karmaker et al., 2021). Governments in emerging economies can benefit businesses by lowering taxes, offering lesser interest loans, and enhancing the incentives plan (Hossain, 2021; Shafi et al., 2020). However, to be sustainable, SCs in the emerging economies need more support, which cannot be given solely by government assistance. As a result, financial aid would be needed from several supply chain partners, depending on their economic power (Karmaker et al., 2021; Metwally et al., 2020; Remko, 2020; Shahin, 2020; A. Sharma et al., 2020, 2021).

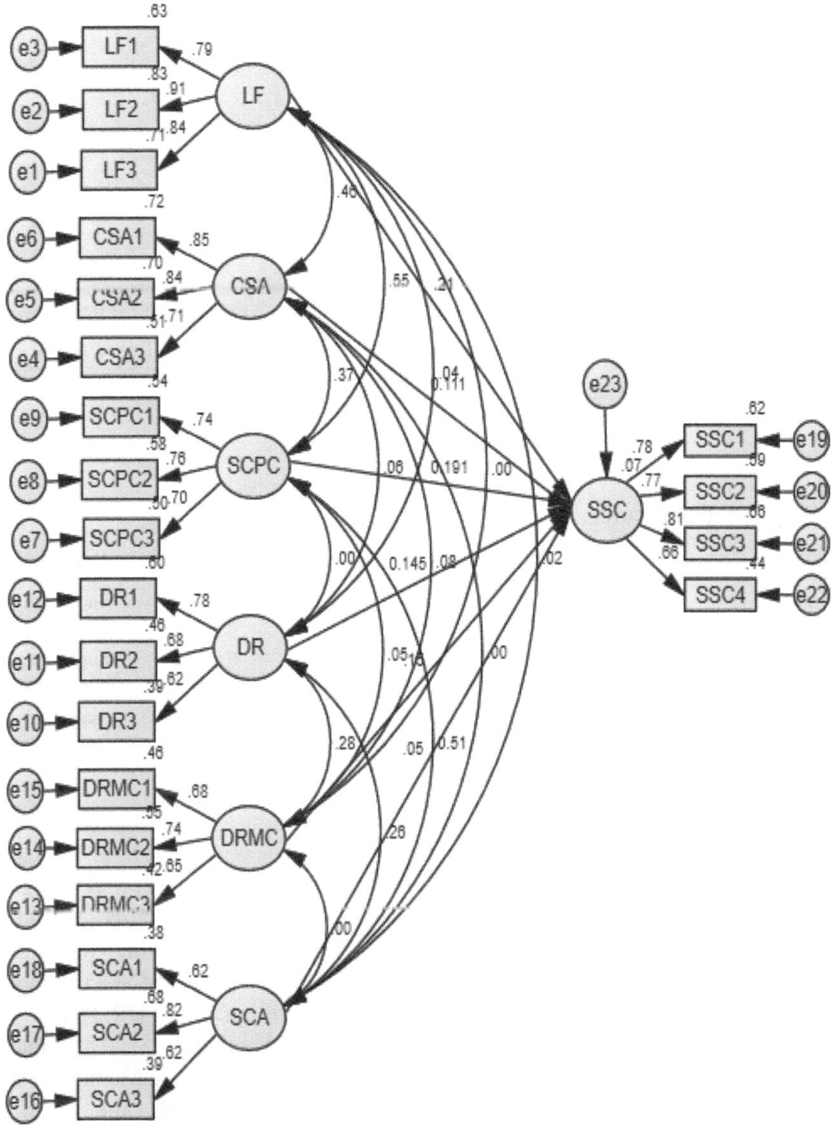

FIGURE 4.2 Final measurement model.

TABLE 4.6
Final Goodness of Fit Indices for the Structural Model

Goodness-of-Fit Indices	Default Model	Benchmark
Absolute goodness-of-fit measure		
CMIN/Df	2.168	⩽3
Absolute badness of fit measure		
RMSEA	0.054	⩽0.08
Incremental fit measure		
CFI	0.933	⩾0.90
IFI	0.934	⩾0.90
TLI	0.918	⩾0.90
Parsimony fit measure		
PCFI	0.759	⩾0.50
PNFI	0.719	⩾0.50

TABLE 4.7
Structural Model Results

	Estimate	SE	CR	P	Hypothesis
SSC<---LF	0.209	0.059	3.54	0.000	Supported
SSC<---CSA	0.111	0.064	1.73	0.000	Supported
SSC<---SCPC	0.191	0.112	1.70	0.000	Supported
SSC<---DR	0.145	0.119	1.21	0.002	Supported
SSC<---DRMC	0.149	0.094	1.58	0.000	Supported
SSC<---SCA	0.51	0.080	6.375	0.000	Supported

The component LF: Strong Legislation Facility to Tackle CV-19 has three indicators – LF1, LF2, and LF3 with values 0.844, 0.929, and 0.902, which show very high loadings (>|.40|); CSA: Customer Support Awareness has three indicators – CSA1, CSA2, and CSA3 with values 0.836, 0.950, and 0.811, which show very high loadings (>|.40|); SCPC: Supplier Side has three indicators – SCPC1, SCPC2, and SCPC3 with values 0.770, 0.874, and 0.809, which show very high loadings (>|.40|); DR: Delivery Reliability has three indicators – DR1, DR2, and DR3 with values 0.731, 0.855, and 0.820, which show very high loadings (>|.40|); DRMC: Disruption Risk Management Capacity has three indicators – DRMC1, DRMC2, and DRMC3 with values 0.840, 0.804, and 0.766, which show very high loadings (>|.40|); SCA: Supply Chain Agility has three indicators – SCA1, SCA2, and SCA3 with values 0.825, 0.866, and 0.697, which show very high loadings (>|.40|). One dependent variable is SSC: Sustainable Supply Chain, with four indicators – SSC1, SSC2, SSC3, and SSC4.

The hypotheses H1, H2, H3, H4, H5, and H6 could not be rejected. Furthermore, LF and SSC have been supported for the structural final model ($\beta = .209$, $p = .000$); support for CSA and SSC ($\beta = .111$, $p = .000$); support for SCPC and SSC ($\beta = .191$, p = .000); support for DR and SSC (β = .145, p = .000); support for DRMC and SSC (β = .194; p = .000); and support for SSC and SSC (β = .451, p = .000) in the current study. In this research, the components are explained and valid with the SEM approach's help, which is the most appropriate method to prove the validity. This technique has not been used until the date in any prior research, making it a unique study.

The business environment and the competition in the market have changed a great deal due to the pandemic (Aday & Aday, 2020; Batra, 2020; Ivanov, 2020b; Prasetyo et al., 2020). Companies need to respond very quickly and make their SC agile. Top management must make the working environment safer for the employees and follow all the pandemic guidelines (Al-Zabidi et al., 2021; Karmaker et al., 2021). To be prepared for future upcoming pandemics, businesses must rethink SC policy growth. Organizations can also inform staff on CV-19 signs and avoidance and appropriate health procedures that impact the SSC (Biswas & Das, 2020; Butu et al., 2020; Cohen, 2020). The "next standard" transition would necessitate increased collaboration between governments and businesses and their competitors and partners to ensure global SC sustainability. Companies need to focus on integrating and working with the SC stakeholders and the respective countries' governments, which is vital for achieving shared goals and handling losses and future progress (Hall et al., 2020; Işık et al., 2021; Pantano et al., 2020). Financial support from the government, such as financial packages and grants for infrastructure development, helps the company survive and maintain average production (Roggeveen & Sethuraman, 2020). Financial funding from SC stakeholders, on the other hand, encourages sustainability-related growth initiatives and facilitates business collaborations with economic benefits in the sense of CV-19 (Brydges et al., 2020; M & Kannappan, 2020; Mahendra Dev & Sengupta, 2020; Sarkis, 2020).

4.7 MANAGERIAL IMPLICATIONS

This study aims to determine the strategies that can help the industries recover from the pandemic's current situations and developing an SSC. If the strategy suggested in this paper is implemented in the sectors by the managers, then they can survive and emerge from this situation. Managers of these firms need to take the lead in implementing these strategies as they will be helpful in the long run. Organizations need to rethink their SC policy and be prepared for pandemic situations in the future. Organizations need to educate employees about CV-19 prevention and symptoms and encourage vaccination. Organizations need to have integration and collaboration with the partners in the SC.

4.8 CONCLUSION

The CV-19 pandemic has given opportunities to various industries to revive their SC. This study aims to identify the strategies and develop a model to help the SC of multiple sectors. For this study, structured literature review is done to determine the strategy for SSC. Six strategies were identified for an SSC. For survey-based research in five industries, a questionnaire was developed. The employees in these companies are the target population. Six hypotheses with six different variables and one dependent variable were developed for this study. The data were analyzed using the analysis of exploratory factors and the modeling of structural equations. A model was developed that meets all parameters and accepts all hypotheses. This strategy is being implemented in the industry and helps SCs to escape the pandemic.

REFERENCES

Aday, S., & Aday, M. S. (2020). Impact of COVID-19 on the food supply chain. *Food Quality and Safety*, *4*(4), 167–180. https://doi.org/10.1093/fqsafe/fyaa024

Adivar, B., Hüseyinoğlu, I. Ö. Y., & Christopher, M. (2019). A quantitative performance management framework for assessing omnichannel retail supply chains. *Journal of Retailing and Consumer Services*, *48*, 257–269. https://doi.org/10.1016/j.jretconser.2019.02.024

Al-Zabidi, A., Rehman, A. U., & Alkahtani, M. (2021). An approach to assess sustainable supply chain agility for a manufacturing organization. *Sustainability*, *13*(4), 1752. https://doi.org/10.3390/su13041752

Amalia, S., Darma, D. C., & Maria, S. (2020). Supply chain management and the Covid-19 outbreak: Optimizing its role for Indonesia. *Current Research Journal of Social Sciences and Humanities*, *3*(2), 196–202. https://doi.org/10.12944/crjssh.3.2.07

Annosi, M. C., Brunetta, F., Bimbo, F., & Kostoula, M. (2021). Digitalization within food supply chains to prevent food waste. Drivers, barriers and collaboration practices. *Industrial Marketing Management*, *93*, 208–220. https://doi.org/10.1016/j.indmarman.2021.01.005

Baral, M. M., Singh, R. K., & Kazançoğlu, Y. (2021). Analysis of factors impacting survivability of sustainable supply chain during COVID-19 pandemic: An empirical study in the context of SMEs. *The International Journal of Logistics Management. https://doi.org/10.1108/IJLM-04-2021-0198*

Baral, M. M., & Verma, A. (2021). Cloud computing adoption for healthcare: An empirical study using SEM approach. *FIIB Business Review*, *10*(3), 255–275. https://doi.org/10.1177/2319714521101 2505

Barman, A., Das, R., & De, P. K. (2021). Impact of COVID-19 in food supply chain: Disruptions and recovery strategy. *Current Research in Behavioral Sciences*, *2*, 100017. https://doi.org/10.1016/j.crbeha.2021.100017

Batra, D. (2020). The impact of the COVID-19 on organizational and information systems agility. *Information Systems Management*, *37*(4), 361–365. https://doi.org/10.1080/10580530.2020.1821843

Biswas, T. K., & Das, M. C. (2020). Selection of the barriers of supply chain management in Indian manufacturing sectors due to Covid-19 impacts. *Operational Research in Engineering Sciences: Theory and Applications*, *3*(3), 1–12. https://doi.org/10.31181/oresta2030301b

Bochtis, D., Benos, L., Lampridi, M., Marinoudi, V., Pearson, S., & Sørensen, C. G. (2020). Agricultural workforce crisis in light of the COVID-19 pandemic. *Sustainability*, *12*(19), 8212. https://doi.org/10.3390/su12198212

Brydges, T., Heinze, L., Retamal, M., & Henninger, C. E. (2020). Platforms and the pandemic: A case study of fashion rental platforms during COVID-19. *Geographical Journal*. https://doi.org/10.1111/geoj.12366

Buhalis, D., & Leung, R. (2018). Smart hospitality – Interconnectivity and interoperability towards an ecosystem. *International Journal of Hospitality Management*, *71*(March 2017), 41–50. https://doi.org/10.1016/j.ijhm.2017.11.011

Butu, A., Brumă, I. S., Tanasă, L., Rodino, S., Vasiliu, C. D., Doboş, S., & Butu, M. (2020). The impact of COVID-19 crisis upon the consumer buying behaviour of fresh vegetables directly from local producers. Case study: The quarantined area of Suceava County,

Romania. *International Journal of Environmental Research and Public Health*, *17*(15), 1–25. https://doi.org/10.3390/ijerph17155485

Byrne, B. M. (2010). *Structural equation modeling with AMOS: Basic concepts, applications, and programming* (2nd edition). New York: Routledge Academy.

Cariappa, A. G. A., Acharya, K. K., Adhav, C., R, S., & Ramasundaram, P. (2020). Pandemic led food price anomalies and supply chain disruption: Evidence from COVID-19 incidence in India. *SSRN Electronic Journal*. https://doi.org/10.2139/ssrn.3680634

Chowdhury, M. T., Sarkar, A., Saha, P. K., & Anik, R. H. (2020). Enhancing supply resilience in the COVID-19 pandemic: A case study on beauty and personal care retailers. *Modern Supply Chain Research and Applications*, *2*(3), 143–159. https://doi.org/10.1108/mscra-07-2020-0018

Cohen, M. J. (2020). Does the COVID-19 outbreak mark the onset of a sustainable consumption transition? *Sustainability: Science, Practice, and Policy*, *16*(1), 1–3. Taylor and Francis Inc. https://doi.org/10.1080/15487733.2020.1740472

DeVellis, R. F., Lewis, M. A., & Sterba, K. R. (2003). Interpersonal emotional processes in adjustment to chronic illness. *Social psychological foundations of health and illness*, 256–287, Editors: Jerry Suls, Kenneth A. Wallston, Blackwell Publishing Ltd, 350 Main Street, Malden, MA 02148-5018, USA.

Fink, L. (2020). Conducting information systems research in the midst of the COVID-19 pandemic: Opportunities and challenges. *Information Systems Management*, *37*(4), 256–259. https://doi.org/10.1080/10580530.2020.1814460

Gössling, S., Scott, D., & Hall, C. M. (2021). Pandemics, tourism and global change: A rapid assessment of COVID-19. *Journal of Sustainable Tourism*, *29*(1), 1–20. https://doi.org/10.1080/09669582.2020.1758708

Gray, R. S. (2020). Agriculture, transportation, and the COVID-19 crisis. *Canadian Journal of Agricultural Economics/Revue Canadienne d'agroeconomie*, *68*(2), 239–243. https://doi.org/10.1111/cjag.12235

Hair, J. F., Black, W. C., Babin, B. J., & Anderson, R. E. (2010). *Multivariate data analysis*. Pearson, NJ: Pearson Education Inc.

Hair, J. F., Ringle, C. M., & Sarstedt, M. (2013). Partial least squares structural equation modeling: Rigorous applications, better results and higher acceptance. *Long Range Planning*, *46*(1–2), 1–12.

Hall, M. C., Prayag, G., Fieger, P., & Dyason, D. (2020). Beyond panic buying: Consumption displacement and COVID-19. *Journal of Service Management*. https://doi.org/10.1108/JOSM-05-2020-0151

Henseler, J., Ringle, C. M., & Sinkovics, R. R. (2009). The use of partial least squares path modeling in international marketing. *New challenges to international marketing, 277-319.* Editors: Rudolf R. Sinkovics, Pervez N. Ghauri, Bingley: Emerald Group Publishing Limited.

Hobbs, J. E. (2020). Food supply chains during the COVID-19 pandemic. *Canadian Journal of Agricultural Economics*, *68*(2), 171–176. https://doi.org/10.1111/cjag.12237

Hossain, M. (2021). The effect of the Covid-19 on sharing economy activities. *Journal of Cleaner Production*, *280*, 124782. https://doi.org/10.1016/j.jclepro.2020.124782

Işık, S., İbiş, H., & Gulseven, O. (2021). The impact of the COVID-19 pandemic on Amazon's business. *SSRN Electronic Journal*. https://doi.org/10.2139/ssrn.3766333

Ivanov, D. (2020a). Predicting the impacts of epidemic outbreaks on global supply chains: A simulation-based analysis on the coronavirus outbreak (COVID-19/SARS-CoV-2) case. *Transportation Research Part E: Logistics and Transportation Review*, *136*(March), 101922. https://doi.org/10.1016/j.tre.2020.101922

Ivanov, D. (2020b). Viable supply chain model: Integrating agility, resilience and sustainability perspectives – lessons from and thinking beyond the COVID-19 pandemic. *Annals of Operations Research*. https://doi.org/10.1007/s10479-020-03640-6

Ivanov, D., & Das, A. (2020). Coronavirus (COVID-19/SARS-CoV-2) and supply chain resilience: A research note. *International Journal of Integrated Supply Management, 13*(1), 90–102. https://doi.org/10.1504/IJISM.2020.107780

Kalogiannidis, S., & Melfou, K. (2020). Issues and opportunities for agriculture sector during global pandemic. *International Journal of Economics, Business and Management Research, 4*(12).

Karmaker, C. L., Ahmed, T., Ahmed, S., Ali, S. M., Moktadir, M. A., & Kabir, G. (2021). Improving supply chain sustainability in the context of COVID-19 pandemic in an emerging economy: Exploring drivers using an integrated model. *Sustainable Production and Consumption, 26*, 411–427. https://doi.org/10.1016/j.spc.2020.09.019

Kline, R. B. (2015). *Principles and practice of structural equation modeling.* Guilford Publications, 370 Seventh Avenue, Suite 1200, New York, NY 10001-1020

Kumar, P., Singh, S. S., Pandey, A. K., Singh, R. K., Srivastava, P. K., Kumar, M., Dubey, S. K., Sah, U., Nandan, R., Singh, S. K., Agrawal, P., Kushwaha, A., Rani, M., Biswas, J. K., & Drews, M. (2021). Multi-level impacts of the COVID-19 lockdown on agricultural systems in India: The case of Uttar Pradesh. *Agricultural Systems, 187*, 103027. https://doi.org/10.1016/j.agsy.2020.103027

M, A. S. R., & Kannappan, S. (2020). Marketing agility and E-Commerce agility in the light of COVID-19 pandemic: A study with reference to fast fashion brands. *Asian Journal of Interdisciplinary Research, 3*(4), 1–13. https://doi.org/10.34256/ajir2041

Mahendra Dev, S., & Sengupta, R. (2020). *Impact of Covid-19 on the Indian economy: An interim assessment.* https://time.com/5818819/imf-coronavirus-economic-collapse/

Mangla, S. K., Luthra, S., Rich, N., Kumar, D., Rana, N. P., & Dwivedi, Y. K. (2018). Enablers to implement sustainable initiatives in agri-food supply chains. *International Journal of Production Economics, 203*(April 2017), 379–393. https://doi.org/10.1016/j.ijpe.2018.07.012

Metwally, A. B. M., Ali, S. A. M., & Mohamed, A. T. I. (2020, October 26). Resilience and agility as indispensable conditions for sustaining viable supply chain during pandemics: The case of Bahrain. *2020 International Conference on Data Analytics for Business and Industry: Way Towards a Sustainable Economy, ICDABI 2020.* https://doi.org/10.1109/ICDABI51230.2020.9325609

Mukherjee, S., & Chittipaka, V. (2021). Analysing the adoption of intelligent agent technology in food supply chain management: An empirical evidence. *FIIB Business Review*, https://doi.org/10.1177/23197145211059243.

Nakat, Z., & Bou-Mitri, C. (2021). COVID-19 and the food industry: Readiness assessment. *Food Control, 121*, 107661. https://doi.org/10.1016/j.foodcont.2020.107661

Nunnally, J. C. (1994). *Psychometric theory 3E.* Tata McGraw-Hill Education, New Delhi, India.

Pal, S. K., Baral, M. M., Mukherjee, S., Venkataiah, C., & Jana, B. (2021). Analyzing the impact of supply chain innovation as a mediator for healthcare firms' performance. *Materials Today: Proceedings.* https://doi.org/10.1016/j.matpr.2021.10.173

Pantano, E., Pizzi, G., Scarpi, D., & Dennis, C. (2020). Competing during a pandemic? Retailers' ups and downs during the COVID-19 outbreak. *Journal of Business Research, 116*, 209–213. https://doi.org/10.1016/j.jbusres.2020.05.036

Podsakoff, P. M., MacKenzie, S. B., Lee, J. Y., & Podsakoff, N. P. (2003). Common method biases in behavioural research: A critical review of the literature and recommended remedies. *Journal of Applied Psychology, 88*(5), 879.

Prasetyo, Y. T., Castillo, A. M., Salonga, L. J., Sia, J. A., & Seneta, J. A. (2020). Factors affecting perceived effectiveness of COVID-19 prevention measures among Filipinos during Enhanced Community Quarantine in Luzon, Philippines: Integrating protection motivation theory and extended theory of planned behaviour. *International Journal of Infectious Diseases, 99*, 312–323. https://doi.org/10.1016/j.ijid.2020.07.074

Qingbin, Wang., Liu, C.-quan, Zhao, Y.-feng, Kitsos, A., Cannella, M., Wang, S.-kun, & Han, L. (2020). Impacts of the COVID-19 pandemic on the dairy industry: Lessons from China and the United States and policy implications. *Journal of Integrative Agriculture, 19*(12), 2903–2915. https://doi.org/10.1016/S2095-3119(20)63443-8

Queiroz, M. M., Ivanov, D., Dolgui, A., & Fosso Wamba, S. (2020). Impacts of epidemic outbreaks on supply chains: Mapping a research agenda amid the COVID-19 pandemic through a structured literature review. In *Annals of Operations Research* (Issue 0123456789). Springer US. https://doi.org/10.1007/s10479-020-03685-7

Raut, R. D., Narkhede, B., & Gardas, B. B. (2017). To identify the critical success factors of sustainable supply chain management practices in the context of oil and gas industries: ISM approach. *Renewable and Sustainable Energy Reviews, 68*, 33–47. Elsevier Ltd. https://doi.org/10.1016/j.rser.2016.09.067

Remko, van H. (2020). Research opportunities for a more resilient post-COVID-19 supply chain – closing the gap between research findings and industry practice. *International Journal of Operations and Production Management, 40*(4), 341 355. https://doi.org/10.1108/IJOPM-03-2020-0165

Richards, T. J., & Rickard, B. (2020). COVID-19 impact on fruit and vegetable markets. *Canadian Journal of Agricultural Economics, 68*(2), 189–194. https://doi.org/10.1111/cjag.12231

Roggeveen, A. L., & Sethuraman, R. (2020). How the COVID-19 pandemic may change the world of retailing. *Journal of Retailing, 96*(2), 169–171. Elsevier Ltd. https://doi.org/10.1016/j.jretai.2020.04.002

Rowan, N. J., & Galanakis, C. M. (2020). Unlocking challenges and opportunities presented by COVID-19 pandemic for cross-cutting disruption in agri-food and green deal innovations: Quo Vadis? *Science of the Total Environment, 748*, 141362). Elsevier BV https://doi.org/10.1016/j.scitotenv.2020.141362

Sajjad, A., Eweje, G., & Tappin, D. (2020). Managerial perspectives on drivers for and barriers to sustainable supply chain management implementation: Evidence from New Zealand. *Business Strategy and the Environment, 29*(2), 592–604. https://doi.org/10.1002/bse.2389

Sarkis, J. (2020). Supply chain sustainability: Learning from the COVID-19 pandemic. *International Journal of Operations and Production Management, 41*(1), 63–73. https://doi.org/10.1108/IJOPM-08-2020-0568

Sarkis, J., Cohen, M. J., Dewick, P., & Schröder, P. (2020). A brave new world: Lessons from the COVID-19 pandemic for transitioning to sustainable supply and production. *Resources, Conservation and Recycling, 159*, 104894. Elsevier BV. https://doi.org/10.1016/j.resconrec.2020.104894

Savary, S., Akter, S., Almekinders, C., Harris, J., Korsten, L., Rötter, R., Waddington, S., & Watson, D. (2020). Mapping disruption and resilience mechanisms in food systems. *Food Security, 12*(4), 695–717. https://doi.org/10.1007/s12571-020-01093-0

Schniederjans, D. G., Curado, C., & Khalajhedayati, M. (2020). Supply chain digitisation trends: An integration of knowledge management. *International Journal of Production Economics, 220*, 107439. https://doi.org/10.1016/j.ijpe.2019.07.012

Shafi, M., Liu, J., & Ren, W. (2020). Impact of COVID-19 pandemic on micro, small, and medium-sized Enterprises operating in Pakistan. *Research in Globalization, 2*, 100018. https://doi.org/10.1016/j.resglo.2020.100018

Shahed, K. S., Azeem, A., Ali, S. M., & Moktadir, M. A. (2021). A supply chain disruption risk mitigation model to manage COVID-19 pandemic risk. *Environmental Science and Pollution Research*, 1–16. https://doi.org/10.1007/s11356-020-12289-4

Shahin, A. (2020). *Supply chain risk management under covid-19: A review and research agenda. October*, 1–24. https://www.researchgate.net/publication/343852256

Shanmugam, K., Jeganathan, K., Mohamed Basheer, M. S., Mohamed Firthows, M. A., & Jayakody, A. (2020). Impact of business intelligence on business performance of food delivery platforms in Sri Lanka. *Global Journal of Management and Business Research*, *20*(6), 39–51. https://doi.org/10.34257/gjmbrgvol20is6pg39

Sharma, A., Adhikary, A., & Borah, S. B. (2020). Covid-19's impact on supply chain decisions: Strategic insights from NASDAQ 100 firms using Twitter data. *Journal of Business Research*, *117*, 443–449. https://doi.org/10.1016/j.jbusres.2020.05.035

Sharma, A., Borah, S. B., & Moses, A. C. (2021). Responses to COVID-19: The role of governance, healthcare infrastructure, and learning from past pandemics. *Journal of Business Research*, *122*, 597–607. https://doi.org/10.1016/j.jbusres.2020.09.011

Sharma, M., Luthra, S., Joshi, S., & Kumar, A. (2021). Accelerating retail supply chain performance against pandemic disruption: Adopting resilient strategies to mitigate the long-term effects. *Journal of Enterprise Information Management* (ahead-of-print). https://doi.org/10.1108/JEIM-07-2020-0286

Sharma, R., Shishodia, A., Kamble, S., Gunasekaran, A., & Belhadi, A. (2020). Agriculture supply chain risks and COVID-19: Mitigation strategies and implications for the practitioners. *International Journal of Logistics Research and Applications*, *0*(0), 1–27. https://doi.org/10.1080/13675567.2020.1830049

Singh, S., Kumar, R., Panchal, R., & Tiwari, M. K. (2020). Impact of COVID-19 on logistics systems and disruptions in food supply chain. *International Journal of Production Research*, *0*(0), 1–16. https://doi.org/10.1080/00207543.2020.1792000

Trotter, P., Mugisha, M. B., Mgugu-Mhene, A. T., Batidzirai, B., Jani, A. R., & Renaldi, R. (2020). Between collapse and resilience: Emerging empirical evidence of COVID-19 impact on food security in Uganda and Zimbabwe. *SSRN Electronic Journal*. https://doi.org/10.2139/ssrn.3657484

Xu, Z., Elomri, A., Kerbache, L., & El Omri, A. (2020). Impacts of COVID-19 on global supply chains: Facts and perspectives. *IEEE Engineering Management Review*, *48*(3), 153–166. https://doi.org/10.1109/EMR.2020.3018420

5 Circular Economy Measures to Diminish the Perils of COVID-19 Outbreak in Aegis of Supply Chain

Somesh Agarwal, Mohit Tyagi, and R.K. Garg

CONTENTS

5.1 INTRODUCTION

Pandemic outbreaks have always had a major impact on society's lifestyle and work culture. Whenever a pandemic or epidemic outbreak had occurred in the past, it had always harmed the organization's working environment and society's well-being. A new pandemic of the SARS-CoV-2 (COVID-19) virus wreaked devastation in 2020, touching nearly every part of the world. Pandemic circumstances have caused governments worldwide to take lockdown actions to stop the virus from spreading, which included shutting down businesses and workplaces and restricting people's movement. This action turned out to be a reason for the economic slowdown and reduced GDP.

DOI: 10.1201/9781003150084-5

Manufacturing industries that are considered the backbone of any country's economy have been affected greatly by this outbreak. Interruptions in the supply chain, unavailability of the workforce, industrial shut down, and troubles in transportation modes had affected the manufacturing sector practices very badly (Ivanov, 2020). The global outbreak of COVID-19 has uncovered significant weaknesses of the contemporary global manufacturing and supply network. The unexpected economic breakdown has shed light on organizations to think differently about the system. The shattering of traditional economic patterns has led the human race to a new horizon, where a stable and robust economy is urgent.

It is proposed that in the future, the circular economy could make enterprises resilient to such pandemic disorders, in particular by developing local manufacturing and supply networks. The circular economy is also promoted as a post-pandemic economic revitalization option that is environmentally responsible.

The circular economy's main idea depicts a production and consumption system focused on item recycling, reusing, refurbishing, remanufacturing, sharing and shifting consumer behaviour, and building new business models and systems (Geissdoerfer et al., 2017). CE is a multidisciplinary framework that necessitates integrating all system-related and non-system-related activities such as production, manufacturing, and consumption. CE aspires to rescue humanity from numerous environmental risks by decreasing resource waste and overuse.

In the world, only 8.6% of total waste has been recycled and reused, but due to the pandemic, this figure will be shrunken due to the fear of contagion of the waste collector and team who process it. A revenue-generating business model based on trash recovery is what CE refers to as a circular economy business model (CEBM). CEBM is a mix of the CE principles and business model methods that create a safer environment and profit for industry (Geissdoerfer et al., 2018). With its characteristics of the closed loop, narrow loop, slowing, changing shapes or configuration, and disinfecting loops, CEBM is a revolutionary innovation in an existing business model that could be defined as its own business model, enabling industries to minimize overuse and reducing waste and emissions outflow from industries.

This work attempts to locate a backing for the miserable conditions generated due to pandemic outbreak using sector-specific solutions through CEBM. Studies showed that the role of COVID-19 in waste growth, especially of plastic, food, and medical waste, is distressing. The theme of this work expresses how this generated waste could be utilized to support the economic recovery from slowdown and sequentially to control and reduce sudden waste growth by employing CE principles. Waste that could be worth thousands of crores might be utilized by using CEBM to enhance the economic structure and social well-being in the current pandemic era where everyone is dealing with the financial crisis. Furthermore, the perils caused by pandemics would remain to propagate in the near future, and this work attempts to provide remedial measures for some identified business sectors, which would help create a resilient and robust economy in the COVID-19 pandemic scenario. Nine significant business sectors were identified in this research: utilization of agricultural and food waste, plastic waste footprint reduction, construction and demolition industry, fashion industry, industrial symbiosis, unemployment, more dependency on the local supply chain, Industry 4.0, and emphasis on renewable energy resources. This chapter suggests that waste reduction would not be merely helpful, but utilizing the waste for a better purpose would be a more beneficial concept.

5.2 MEASURES TO MITIGATE THE PANDEMIC PERILS THROUGH CEBM

Business models are designed on two fundamental objectives: customer satisfaction and revenue generation. When the concept of CE is assimilated into the business model, then an additional feature of environmental security is added to its existing objectives. CEBM is a business model that works on the philosophy of CE. This model helps to use products or materials to the extent of their life cycle and extract maximum values from them, keeping the value chain's revenues intact. Implementation of CE in the business model gives opportunities for sustainable development, which in turn generates employability opportunities, increases the efficiency of resource utilization, reduces waste, and helps produce new innovative products (Geissdoerfer et al., 2018). Despite the advantages of cost-effectiveness and environmental safety, the CEBM has not been commonly adopted in the consumer market. Delay in the implications of CEBM is due to the customer's reluctant behaviour towards it (Abbey et al., 2015). But coronavirus has changed individuals' mindsets, which has resulted in them focusing on revenue generation, a sustainable environment, and better well-being. This research explores some major business models and attempts to integrate CE practices in these models to emerge from the perils of the current COVID-19 pandemic.

Figure 5.1 shows the identified nine significant business sectors, on which the application of CE is analysed to create a new CEBM for each sector that can aid in

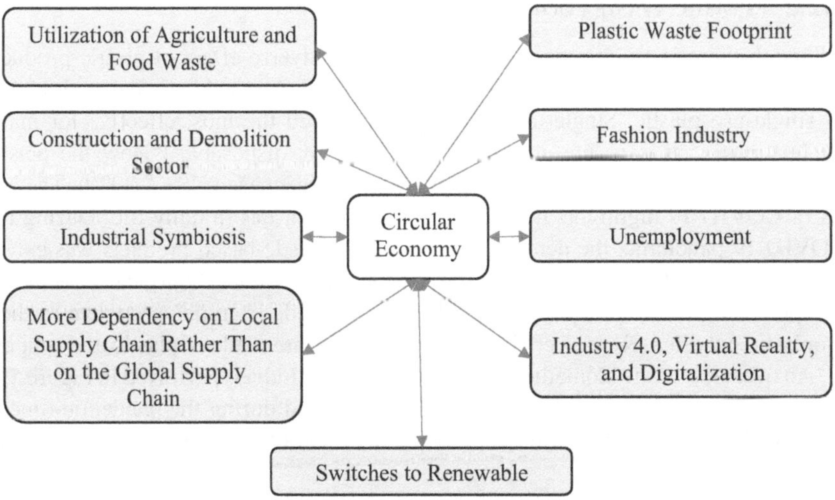

FIGURE 5.1 CE solution to different business sectors to mitigate the perils of the COVID-19 outbreak.

reducing the perils of COVID-19 on the aforementioned sectors. CEBM for each specific business sector has been detailed as follows.

5.2.1 Utilization of Agriculture and Food Waste

The agricultural waste of rotten crops has been escalating because of lockdown and the interruption of the supply chain, resulting in the accumulation of crop remnants. Supply chain disturbance also increases animal residues, which were earlier used for biomass production in big plants (Sharma et al., 2021). In addition to this, kitchen and home food waste also increase abundantly. Rotten crops waste, crop remnants, and food wastage from home could be used for preparing compost. The innovations have led to the scenario where people are not dependent on the big industries to treat these kinds of wastes; they could treat and transform these wastes into compost on their own. This healthy compost is a medicine for crops and beneficial for environmental prosperity (Cristóbal et al., 2018), and it would also prevent the usage of chemical fertilizers. Chemical fertilizers have a very hazardous effect on crops, lands, and their consumption through the final product. Composting could stimulate waste reduction, align with the CE principle, and help people attain a good quality homemade compost, which could be helpful for healthier plants by reducing carbon footprints. The current COVID pandemic encourages people towards herd immunity, for which people are concentrating on the consumption of healthier foods and nutrients for which organically treated farmland and less pesticide usage is the centre of interest.

Additionally, agricultural waste and remnants could be transformed into fibres that could satisfy several functions such as packaging material, disposable food plates, and household items. These products are readily biodegradable and would serve the environment by their zero-carbon emissions capability.

5.2.2 Plastic Waste Footprint Reduction

Although there is an enormous awareness of the adverse effect of plastic products on the environment, the current COVID-19 pandemic has suddenly boosted the use of single-use plastic. Single-use plastic is considered the most effective for many applications as it is a safe alternative and is easily disposable ("How the plastic industry is exploiting anxiety about COVID-19 – Greenpeace USA," 2020). The crisis of COVID-19 highlights the significant role of plastics in daily life. During the COVID-19 pandemic, the demand for single-use plastic-based products was extensively increased for India's medical protection and treatment. Surgical face masks and personal protective equipment (PPE) kits are made from polypropylene, a thermoplastic polymer (Klemeš et al., 2020), and gloves are made of plastic and rubber.

An overview of the biomedical waste generation in India is portrayed in Figure 5.2. The figure reveals that the amount of waste generated during the pandemic time is excessively high, and in the middle of September and October 2020, it is at its highest level.

Contaminated waste is disposed of in single-use plastic bags for better hygienic conditions. Single-use plastic is heavily used for packaging medicines and syringes and keeping contaminated samples and medical-related items. This plastic material

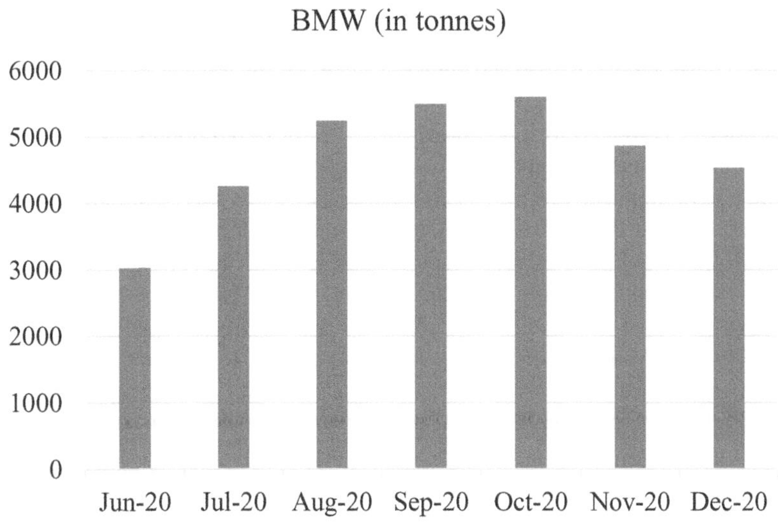

FIGURE 5.2 Biomedical waste generated in India during COVID-19 pandemic (in tonnes). *Source*: Chand et al., 2021

is a waste product after its use and is challenging predominantly because of the need to destroy it to reduce contagion. This has resulted in creating tonnes of plastic waste solely from the medical field (Rowan and Laffey, 2020), which has questioned the practices of environmental welfare. This has generated two significant concerns – improper waste disposal leading to infection in the community and the growth of this disposed waste. Adequate identification, collection, separation, storage, transport, treatment, and disposal, and all related aspects such as disinfection, personal protection, and training are required to manage biomedical and healthcare waste effectively. As there is the fear of infection from medical waste, the items cannot be reused or recycled. Incineration of this waste is a more suitable option combined with waste heat recovery to retrieve the valuable chemical energy content of plastics (Klemeš et al., 2020).

Lockdown actions have also led to an upsurge in the quantity of packaging used for the food and groceries delivered to residences (Kahlert and Bening, 2020). Such transformations would intensify environmental problems with plastics that were already the biggest problem for society even before the pandemic. While this growth is inevitable, attempts should be continued to protect the environment. Recycling techniques could be availed for transforming used plastics to raw material and using modern technologies such as Industry 4.0, which would prevent the workforce from coming into direct contact with the material. Moreover, thermal processes such as burning, autoclaving, steam treatment, and microwave therapy could be used for different treatment techniques. The treatment option depends on multiple factors of acceptability in cultural, technological, environmental, and social issues. Economics, pollution, health, regulatory problems, and public acceptance include related aspects (Yu et al., 2020).

5.2.3 Construction and Demolition Sector

India is a developing country, and construction work always runs at its full pace. But the wastage produced from demolition is massive; sometimes it is reused, but most of the time it is treated as a complete waste. It not only pollutes the environment but also causes inappropriate landfilling. It is desirable to restrict or even prohibit building demolition and ensure that construction methods revolve around restoration and reuse. Due to the COVID-19 pandemic, supply chain disruption, and unavailability of the workforce, the price of raw material has been decreased severely. The vital role of this industry is in providing the raw materials required for a low-carbon economy transition. To meet this need, the sector must be able to transfer its production to new resources, preferably to green materials as a measure of CE, and versatile in response to rapid technological changes, requiring some amount of financial resilience (Laing, 2020). New construction technologies based on sustainable circular principles would be beneficial in the current COVID-19 epidemic era. CE governed green initiatives in the construction sector are a breakthrough in this field (Agarwal et al., 2021a), and implementing this would be beneficial for the environment. Moreover, periodic maintenance of the buildings would lengthen their lives, resulting in waste leakage delay.

5.2.4 Fashion Industry

The fashion/textile industry is one of the leaders among all the industries, generating substantial revenues. It's estimated to be worth Rs 108 billion in 2019 in India (Crewe, 2008), employing 45 million people directly in this industry. People are investing millions of crores in clothing purchases, and the model of its usage is linear, that is, purchase, use, and waste. Textile products after usage are discarded as waste, which increases existing garbage and creates landfill. This waste, if used in an accurate direction and in a well-managed way, could provide enormous advantages to the people. Waste clothes and accessories could be delivered to needy people to wear, which is the second principle of CE (reuse). For that, a robust supply chain structure is needed. Many social service organizations provide such kinds of services, and a more robust network could create a more efficient system.

Moreover, clothing waste could be used as a raw material by recycling its fibres, fabric, cloths, and other accessories (EMF, 2017), which is the third principle of CE (recycling). This raw material could be used to manufacture new garments and packaging material and could be transformed into new forms as per consumer demand. Due to COVID-19 pandemic conditions, people would not be willing to use older fabrics because of transmission anxiety. Industries could eradicate this fear, however, by appropriately sanitizing and disinfecting the older garments and giving safety assurance to the customer. The benefits include not only providing clothing at a cheaper cost, but also reducing waste and supporting the development of a sustainable environment. Also, cloth packaging material would reduce the consumption and production of plastic-based packaging items. To enhance the purchasing in this COVID era when people are earning less, cheaper clothing material could be a breakthrough.

5.2.5 Industrial Symbiosis

Industrial interrelation towards achieving waste reduction as a measure of CE is the urgent need of society. CE emphasizes the industrial symbiosis such that waste of one industry could be the raw material for another (Kerdlap et al., 2019); waste of rubber tires could be used for making shoe soles, for example. The COVID-19 pandemic situation has provided ample time for industries to think in this context. Enhancement of the supply chain structure, reverse supply chain partnership, and transparent profit sharing could be implemented for better industrial symbiosis. Large-scale industries could encourage medium and smaller enterprises (MSME) to engage in a better practice towards robust system development. Industrial symbiosis aims to create loops of technological or biological material while reducing leakages and waste in loops – demonstrating the main components of a circular economy during the current COVID-19 pandemic era. The collaboration of business would help improve the work culture and provide better products and services to the customers in an era where many businesses were shut down due to the pandemic. Industrial symbiosis additionally stimulates new job openings, which is necessary for the current crisis condition.

According to a report published by *The Financial Express* on 11 May 2020, constructive effects of industrial symbiosis have been shown by automobile companies who help the people in this crisis condition by making available medical equipment in short supply. In collaboration with different industries initiated by the appeal of government, automobile industries started producing masks, PPE kits, and life-saving ventilators. This is an absolute live example of CE by which resources with automobile sectors are used not for the sole purpose they are meant for but for other purposes for which they are raised. This signals the philosophy of CE to utilize resources appropriately and efficiently, which has risen only because of the crisis compelled by COVID-19. Additionally, this example shows that even if the international system is challenged, the economic conditions of the industries can continue to function. This way, CE could offer a business commitment to the industries in the current COVID state.

5.2.6 Unemployment

The world has to face a significant problem of unemployment due to the COVID-19 pandemic. The unemployment rate became very high as a side effect of the COVID-19 pandemic. The Figure 5.3 shows the effect of the pandemic on the nationwide unemployability rate.

According to the Figure 5.3, the unemployment rate in India suddenly rises in April 2020 due to the outbreak, and the effects of this still have not been overcome.

The unemployability rate of the nation should be lowered as soon as possible, and effort should be made to maintain it as low as possible. This would support the growth of the nation and also enhances the country's GDP. Education and training towards CE could open doors for new mindsets to work in this field (Haigh and Baunker, 2020). The CE promotes workers with the ability to implement changes to promote an adaptive, life-long learning culture. Enabling CEBM to industries would

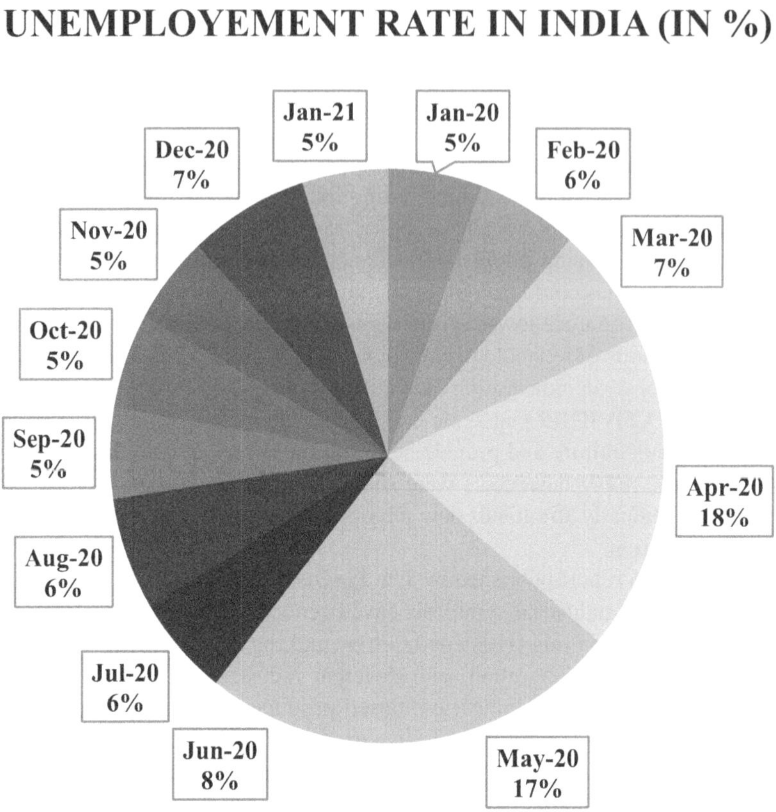

FIGURE 5.3 The unemployment rate in India due to the COVID-19 pandemic outbreak.

Source: Sandhya Keelery, 2020

expand their roles, and for that, they would need additional employees. Processes such as waste collection, processing, recycling, reforming, and transportation would employ the unskilled workforce. Designing and modifying the supply chain structure, reforming the system structure, handling critical responsibilities, operating the system, and so on would generate jobs for skilled workers. It would not be only industries that would generate employability, but also other sectors which are directly or indirectly involved. The government is also opening doors for new mindsets to measure initiatives towards environmental sustainability, which could help individuals be employable and have a high income. Training and learning of CE practices would help the existing employees to work on the required responsibilities and promotions. The situation so created would enable the employers and employees to rethink the way they work as well as how they utilize resources. CE allows the employees and employers to adapt and learn from the market new processes, technologies, and innovation.

There are great opportunities for new and existing micro-industries to recycle and upcycle old products to enrich the environment, which has been a highlighted agenda these days. As the demand for a safer environment is high, any industries that take part in this action would benefit in the present and future. These steps also help in creating jobs and would reduce unemployment rates.

5.2.7 More Dependency on Local Supply Chain Rather Than on the Global Supply Chain

Due to the COVID-19 pandemic, the global supply chain has been disrupted very severely, and people are dependent on local, regional, and national supply chains. the Indian prime minister created a golden slogan on 12 May 2020 – "vocal for local" – while addressing the nation, which not only empowers local suppliers and traders but also decreases the dependency on global markets. To create a self-reliant economy and overcome COVID-19, the Indian prime minister in the same address has taken initiatives towards "Atmanirbhar Bharat" to increase manufacturing and innovations in the industrial sector. This vision decreases the nation's dependency on the global markets, as the objective of CE suggests that population dependency on local/regional suppliers and traders for purchasing would help to circulate the product more effectively within a smaller province (Tyagi et al., 2015). This would enhance the effectiveness of the reverse supply chain by the ease in the collection of waste or used products and ultimately meet the goals of CE (Ethirajan and Kandasamy, 2019). This would also improve the reusing, recycling, and remanufacturing capabilities across the nation and upgrade information sharing between producer and consumer. If this practice continues, the country's GDP would enhance, and the country's overall economy would be improved. Worry towards green initiatives and environmental sustainability would also be decreased due to united working directions.

5.2.8 Industry 4.0, Virtual Reality, and Digitalization

The major failure of the supply chain occurs due to the failure of information flow, which could be overcome by the application of innovation like Industry 4.0. Techniques like the Internet of things (IoT), artificial intelligence (AI), automation, and involvement of robotic technology fabricate innovative industries and support the emerging implementation of CE (Iyer, 2018, Wasiul et al., 2020). The fourth industrial revolution's transformation of traditional manufacturing and industrial practices into modern and intelligent technologies is carried out. The vital principles of Industry 4.0 – decentralization, information readiness, and rapid information exchange networks – could help in achieving augmented sustainable supply chain resolutions (Ethirajan and Kandasamy, 2019). The connection between CE principles and Industry 4.0 practices is still a topic of debate these days. CE facilitates the production of innovative products which are based on the theme of green manufacturing, with a reduction of resource utilization by minimizing waste leakage. Industry 4.0 allows better traceability of smart products across the supply chain; smart technologies enable intelligent maintenance of product during its usage which would decrease the resource utilization. This, in turn, decreases resource consumption and therefore the environmental effects. Integration of Industry 4.0 and CE have motivated business organizations to evolve towards effective and prompt sustainable supply chain management (Lopes de Sousa Jabbour et al., 2018). It also supports the manufacturer in optimizing the product and production methods. In addition to this, it offers an automated solution to a variety of production challenges and meeting the customized demands (Javaid et al., 2020). Enabling Industry 4.0 practices could help maintain social distancing within the industry and make the production easier and faster, which could be controlled from anywhere. Industry 4.0 innovations could lead to better isolation of the COVID-19-infected patient and thereby minimize disease transmission. Modified robotic technologies and sensors might help in hospitals to serve the patients virtually and in public places to watch for overcrowding and to measure the temperature of the people. They could also take and manage samples, serve patients, and possibly disinfect and sterilize to prevent human transmission of the virus (AlMaadeed, 2020). Moreover, 3D printing might be used to manufacture faster medical safety equipment with a shorter supply in the market.

IoT is a network of devices capable of communicating and sensing with embedded intelligent technologies. It offers electronic circuits so designed to integrate various sensors which could measure several parameters. An Internet connection allows data to be collected and exchanged. IoT commonly initiates the phenomenon of the virtual reality (VR) concept. VR is a simulation generated by a computer that creates an artificial 3D environment. The VR concept constitutes input devices, output signals, and graphical interface software. VR would be highly serviceable in current COVID-19 pandemic conditions for learning and training purposes, especially in the medical field for the precise and effective treatment led by experienced doctors' teams. Virtual reality is functional for remote areas to pursue telemedicine to schedule, treat, and monitor diseases (Singh et al., 2020). The virtual reality concept enables a person to be instantly available without having an actual presence. This helps diminish the use of unnecessary resources and reduces worthless travel. Due to the pandemic outbreak, the current teaching and learning process is solely dependent on virtual reality. Enhancing this concept would be beneficial for maintaining social distancing and diminishing travel, which could reduce the infection rates from the deadly coronavirus.

As an aid to a CE, digitization plays a crucial role in helping businesses close loops, implementing more efficient processes, reducing waste, supporting products for longer lives, and lowering their transaction costs (Ivanov, 2020). Digitalization (paperless) reduces person-to-person contact and the transmission of the virus through currency notes and paper usage for information sharing, essential in the current COVID-19 pandemic condition. Digitalization supports restricting unnecessary movements and supply interruption (Tyagi et al., 2014).

5.2.9 Switches to Renewable Energy from Fossil Fuels

India is the third most significant renewable energy producer worldwide (RECAI, 2021). However, the portion of the energy produced through fossils is higher than that of renewable sources. Table 5.1 shows the percentage of the renewable and non-renewable sources used for electricity generation in India.

Table 5.1 demonstrates that the share of non-renewable sources for power generation in India is much higher than that of renewable energy sources. The usage of

TABLE 5.1
Share of Renewable and Non-Renewable Energy Sources for Power Generation in India

Type	Source	% Share
Non-renewable energy	Coal	56.09%
	Gas	6.84%
	Diesel	0.14%
	Subtotal	**63.07%**
Renewable energy	Large hydro plants	12.05%
	Solar power	10.61%
	Wind power	10.59%
	Nuclear power	1.83%
	Other	1.92%
	Subtotal	**37.00%**

(*Source*: Ministry of New and Renewable Energy, 2021)

non-renewable sources of energy stresses the natural reservoir and creates land and water pollution.

However, due to nationwide lockdown, which resulted in reduced transportation and shutting down of industries, people felt relaxed and fresh in the breathable air. Emissions due to CO and NO_2 were reduced considerably, which improved the air quality by 60% (Mahato et al., 2020). Indeed, less air pollution reduces the chances of lung diseases, but only for a short while. Now the question is, what if this state of good air quality could be extended?

This could be achieved by increasing the share of renewable energy and green fuels (Agarwal et al., 2021b) in daily lives instead of limited fossil fuels. Shifting towards renewable sources of energy (green energy) such as solar energy, wind energy, hydro energy, and biomass energy would limit environmental degradation and benefit financially. Strict regulation for the industries for low carbon emissions reduces the usage of fossil fuels, and more emphasis on renewable energy resources operations could enable the alternative usage of renewable energy resources.

5.3 CONCLUSION AND FUTURE STRENGTH

CE empowers the circulation of resources in the value chain, reducing the immediate need for virgin raw material. This cuts the unnecessary burden on the supply chain structure outside the country for the production that depends on global raw materials. The present paper aims at reducing the economic perils and waste upsurge due to the COVID-19 pandemic outbreak by using CE practices.

Slowing economic conditions cost workers their jobs and earnings, increasing the country's unemployment rate. CE allows micro-industries to start reusing and recycling outdated products, which improves the environment and reduces unemployment. Industrial symbiosis would benefit all concerned industries. The involvement of the more extensive industrial sector with MSME would make it easier to follow the process of developing zero-emission waste. Expert R&D sectors of larger-scale enterprises are also needed to innovate plastic waste reuse. Plastic trash incineration is polluting and should be halted. Industrial and university investigators are working hard to reuse and recycle plastic trash, but more research is needed in this area. Industry 4.0 procedures could speed up production with little risk of spreading the virus to workers. Using Industry 4.0-enabled robotic technology, the medical profession revolutionizes patient care, infection sampling, disinfection, and sterilization. Manufacturing with zero-waste leakage is possible with Industry 4.0 and CE practices working together. IoT, a key component of Industry 4.0, facilitates supply chain management through rapid information sharing.

On a concluding note, the present chapter attempts to empower CE practices by providing solutions to 11 different business sectors by rectifying the limitations of these sectors and encouraging them to implement innovative and modified practices to deal with the current COVID-19 pandemic as well as for the future demand of environmental issues.

Further, this study might be carried out in the form of case studies and surveys to track their actual implications. Statistical data could be used to test the performance of the suggested solutions empirically. Also, broader business sectors could be broken into fragments for better investigation of performance impact.

REFERENCES

Abbey, J.D., Meloy, M.G., Guide, V.D.R., Atalay, S., 2015. Remanufactured products in closed-loop supply chains for consumer goods. *Production and Operations Management* 24, 488–503. https://doi.org/10.1111/poms.12238

Agarwal, S., Tyagi, M., Garg, R.K., 2021a. Assessment of barriers of green supply chain management using structural equation modeling, in: *Lecture Notes in Mechanical Engineering.* Springer, Singapore, pp. 441–452. https://doi.org/10.1007/978-981-15-8704-7_55

Agarwal, S., Tyagi, M., Garg, R.K., 2021b. *Commencement of Green Supply Chain Management Barriers: A Case of Rubber Industry,* in: *Lecture Notes in Mechanical Engineering.* Springer, Singapore, pp. 685–699. https://doi.org/10.1007/978-981-15-8542-5_59

AlMaadeed, M.A., 2020. Emergent materials and industry 4.0 contribution toward pandemic diseases such as COVID-19. *Emergent Materials* 3, 107–108. https://doi.org/10.1007/s42247-020-00102-4

Chand, S., Shastry, C.S., Hiremath, S., Joel, J.J., Krishnabhat, C.H., Mateti, U.V., 2021. Updates on biomedical waste management during COVID-19: The Indian scenario. *Clinical Epidemiology and Global Health* https://doi.org/10.1016/j.cegh.2021.100715

Crewe, L., 2008. Global fashion industry. *Spring* 93, 25–33.

Cristóbal, J., Castellani, V., Manfredi, S., Sala, S., 2018. Prioritizing and optimizing sustainable measures for food waste prevention and management. *Waste Management* 72, 3–16. https://doi.org/10.1016/j.wasman.2017.11.007

EMF, 2017. A new textiles economy: Redesigning fashion's future. *Ellen MacArthur Found.* 1–150.

Ethirajan, M., Kandasamy, J., 2019. An analysis on sustainable supply chain for circular economy, in: *Procedia Manufacturing.* Elsevier B.V., pp. 477–484. https://doi.org/10.1016/j.promfg.2019.04.059

Geissdoerfer, M., Morioka, S.N., de Carvalho, M.M., Evans, S., 2018. Business models and supply chains for the circular economy. *Journal of Cleaner Production* 190, 712–721. https://doi.org/10.1016/j.jclepro.2018.04.159

Geissdoerfer, M., Savaget, P., Bocken, N.M.P., Hultink, E.J., 2017. The circular economy – A new sustainability paradigm? *Journal of Cleaner Production* 143, 757–768. https://doi.org/10.1016/j.jclepro.2016.12.048

Haigh, L., Baunker, L., 2020. Covid-19 and the circular economy: Opportunities and reflections [WWW Document]. Medium. URL: https://medium.com/circleeconomy/covid-19-and-the-circular-economy-opportunities-and-reflections-7c2a7db70900 (accessed 6.5.20).

How the plastic industry is exploiting anxiety about COVID-19 – Greenpeace USA [WWW Document], 2020. URL: https://www.greenpeace.org/usa/how-the-plastic-industry-is-exploiting-anxiety-about-covid-19/ (accessed 6.28.20).

Ivanov, D., 2020. Predicting the impacts of epidemic outbreaks on global supply chains: A simulation-based analysis on the coronavirus outbreak (COVID-19/SARS-CoV-2) case. *Transportation Research Part E: Logistics and Transportation Review* 136, 101922. https://doi.org/10.1016/j.tre.2020.101922

Iyer, A., 2018. Moving from Industry 2.0 to Industry 4.0: A case study from India on leap-frogging in smart manufacturing. *Procedia Manufacturing*. 21, 663–670. https://doi.org/10.1016/j.promfg.2018.02.169

Javaid, M., Haleem, A., Vaishya, R., Bahl, S., Suman, R., Vaish, A., 2020. Industry 4.0 technologies and their applications in fighting COVID-19 pandemic. *Diabetes and Metabolic Syndrome: Clinical Research and Reviews* 14, 419–422. https://doi.org/10.1016/j.dsx.2020.04.032

Kahlert, S., Bening, C.R., 2020. Plastics recycling after the global pandemic: Resurgence or regression? *Resources, Conservation and Recycling*. https://doi.org/10.1016/j.resconrec.2020.104948

Kerdlap, P., Low, J.S.C., Ramakrishna, S., 2019. Zero waste manufacturing: A framework and review of technology, research, and implementation barriers for enabling a circular economy transition in Singapore. *Resources, Conservation and Recycling* 151, 104438. https://doi.org/10.1016/j.resconrec.2019.104438

Klemeš, J.J., Fan, Y. Van, Tan, R.R., Jiang, P., 2020. Minimising the present and future plastic waste, energy and environmental footprints related to COVID-19. *Renewable and Sustainable Energy Reviews*. 127. https://doi.org/10.1016/j.rser.2020.109883

Laing, T., 2020. The economic impact of the Coronavirus 2019 (Covid-2019): Implications for the mining industry. *Extractive Industries and Society* 7, 580–582. https://doi.org/10.1016/j.exis.2020.04.003

Lopes de Sousa Jabbour, A.B., Jabbour, C.J.C., Godinho Filho, M., Roubaud, D., 2018. Industry 4.0 and the circular economy: A proposed research agenda and original roadmap for sustainable operations. *Annals of Operations Research* 270, 273–286. https://doi.org/10.1007/s10479-018-2772-8

Mahato, S., Pal, S., Ghosh, K.G., 2020. Effect of lockdown amid COVID-19 pandemic on air quality of the megacity Delhi, India. *Science of the Total Environment* 730, 139086. https://doi.org/10.1016/j.scitotenv.2020.139086

Ministry of New and Renewable Energy, 2021. *Ministry of New and Renewable Energy*. Government of India: Annual report 2020–2021, 53, 1689–1699.

RECAI, 2021. Renewable energy country attractiveness index: Issue 41. *Recai,* 40.

Rowan, N.J., Laffey, J.G., 2020. Challenges and solutions for addressing critical shortage of supply chain for personal and protective equipment (PPE) arising from Coronavirus disease (COVID19) pandemic – Case study from the Republic of Ireland. *Science of the Total Environment* 725, 138532. https://doi.org/10.1016/j.scitotenv.2020.138532

Sandhya Keelery, 2020. India: Unemployment rate due to COVID-19 | Statista [WWW Document]. Statista. URL: https://www.statista.com/statistics/1111487/coronavirus-impact-on-unemployment-rate/ (accessed 7.19.21).

Sharma, J., Tyagi, M., Bhardwaj, A., 2021. Exploration of COVID-19 impact on the dimensions of food safety and security: A perspective of societal issues with relief measures. *Journal of Agribusiness in Developing and Emerging Economies*. https://doi.org/10.1108/JADEE-09-2020-0194

Singh, R.P., Javaid, M., Kataria, R., Tyagi, M., Haleem, A., Suman, R., 2020. Significant applications of virtual reality for COVID-19 pandemic. *Diabetes and Metabolic Syndrome: Clinical Research and Reviews* 14, 661–664. https://doi.org/10.1016/j.dsx.2020.05.011

The Financial Express, 2020. How India's automotive sector turned saviours during Covid-19 scare: Making ventilators, masks, PPE and more. *The Financial Express* [WWW Document]. URL: https://www.financialexpress.com/auto/industry/how-indias-automototive-sector-turned-saviours-during-covid-19-scare-making-ventilators-masks-ppe-hyundai-maruti-ashok-leyland/1949819/ (accessed 7.25.20).

Tyagi, M., Kumar, P., Kumar, D., 2014. Selecting alternatives for improvement in IT enabled supply chain performance. *International Journal of Procurement Management* 7, 168–182. https://doi.org/10.1504/IJPM.2014.059553

Tyagi, M., Kumar, P., Kumar, D., 2015. Assessment of critical enablers for flexible supply chain performance measurement system using fuzzy DEMATEL approach. *Global Journal of Flexible Systems Management* 16, 115–132. https://doi.org/10.1007/s40171-014-0085-6

Wasiul, S., Rizvi, H., Agrawal, S., Murtaza, Q., 2020. Circular economy under the impact of IT tools: A content-based review. *International Journal of Sustainable Engineering* 00, 1–11. https://doi.org/10.1080/19397038.2020.1773567

Yu, H., Sun, X., Solvang, W.D., Zhao, X., 2020. Reverse logistics network design for effective management of medical waste in epidemic outbreaks: Insights from the coronavirus disease 2019 (COVID-19) outbreak in Wuhan (China). *International Journal of Environmental Research and Public Health* 17, 1770. https://doi.org/10.3390/ijerph17051770

6 Deteriorating Inventory Policy in a Two-Warehouse System under Demand Disruption

Achieving Sustainability under COVID-19 Pandemic

Ranveer Singh Rana, Leopoldo Eduardo Cárdenas-Barrón, Harshit Katurka, and Dinesh Kumar

CONTENTS

6.1 INTRODUCTION

It has been a year since the wake of the COVID-19 pandemic. All the production and business activities had been put to a halt for around three months, leading to heavy losses and economic distress of the nation. As the first wave has ended, manufacturing, production, and transportation of goods have started again, and business has resumed. However, as we were all coping, covering our losses, and building back

DOI: 10.1201/9781003150084-6

our economy, we are now being attacked by the new COVID-19 strain in the form of a second wave. With an already fallen economy trying to pull itself together and people suffering from the pandemic's first impact, the second wave situation is like rubbing salt in the wound. We should be prepared to lessen the influence of such pandemics on society because a further spread of viruses is expected (Donthu and Gustafsson, 2020).

As the lockdown was enforced, movements were restricted to slow down the spread of the virus, and governments had to make many decisions that were painful for the common people. These measures proved effective in curbing the spread of the virus, but these regulations also impacted our economy. The disruption caused by these measures in the food supply chain was discussed by Mahajan and Tomar (2021), whereas Wang et al. (2020) discussed its impact on the agriculture sector. Al-Mansour and Al-Ajmi (2020) insisted that business organizations should replan their strategies to combat the COVID-19 situation. After that, Shahed et al. (2021) discussed measures to lessen the influence of natural disasters on any supply chain. Though containment measures disturbed almost all businesses, a positive impact on e-business has been observed and discussed by some researchers as follows. According to Al-Omoush et al. (2020), the coronavirus forced business establishments to adopt e-business to survive in the market. Qian (2020) also mentioned that growth is registered in e-commerce due to COVID-19. Motives of online shopping are discussed by Koch et al. (2020) during COVID-19.

Government rules are very strict regarding carbon emissions, and people are more conscious about the environment. Therefore, the reduction of carbon emissions is a crucial area of interest for researchers. One of the first models of a green supply chain was given by Diabat and Simchi-Levi (2009). Another essential model was proposed by Hua et al. (2011). They presented the influence of carbon trade and carbon cap on the total cost.

Inventory management of deteriorating items is a big challenge for retailers. Its quantity decreases due to many reasons like spoilage, evaporation, damage, and expiration. Vegetables, packed food, medicine, and blood are a few examples of deteriorating items. The first model where deterioration of product is considered was given by Ghare and Schrader (1963). Since then, many researchers have reviewed product deterioration in their research studies (Chakrabarty et al. 1998; Mondal et al. 2003; Shaikh et al. 2017; (Rana et al. 2021).

In this chapter, many real-world phenomena such as supply chain disruption, carbon emissions reduction, deterioration of the product, and time-varying demand rate are considered when the retailer uses two warehouse systems to satisfy customer demand. A numerical experiment is presented to show the influence of demand disruption and the duration of lockdown.

The remaining part of the chapter is arranged as follows: Related literature is given in Section 6.2. The need for the inventory model with disrupted demand, along with notations and assumptions, are mentioned in Section 6.3. Mathematical modeling and analysis are discussed in Section 6.4. The solution procedure is discussed in Section 6.5. In Section 6.6, a numerical example is given. Section 6.7 presents sensitivity analysis. Further, in Section 6.8, the managerial implications are given. Finally, conclusion and future scope are mentioned at the end of this chapter.

6.2 LITERATURE REVIEW

Some literature related to disruption, carbon emissions, and deterioration of items are provided in this section.

6.2.1 Supply Chain Disruption

In 2020–20121, the world has witnessed waves of COVID-19, the most significant supply chain disruptive event. With around 21 million corona cases and more than 200,000 deaths, the pandemic indeed has taken a lot from us. With patients and deaths nearly quadrupled, the second wave has proven to be more dreadful and dangerous than its predecessor. Messina et al. (2020) discussed methodologies used by decision makers to deal with disruptions. According to Grida et al. (2020), the government has taken various containment measures to stop the spread of viruses that restricted movement, which affected almost all the supply chain stages. After that, Mahajan and Tomar (2021) discussed the disruption caused in the food supply chain due to lockdown imposed to stop the spread of the virus. Afterward, Shahed et al. (2021) proposed a model to lessen the effect of disruption in a supply chain network which consists of three stages: supplier, manufacturer, and retailer. Any disruptive event that starts from a point and propagates through the stages is called a ripple effect. Many researchers have mentioned the influence of the ripple effect on supply chain performance and resilience, such as Dolgui and Ivanov (2021), Li and Zobel (2020), Li et al. (2021), and Birkie and Trucco (2020). Measures need to be taken by the food supply chain post-COVID for revival are mentioned by Mor et al. (2020).

6.2.2 Carbon Emissions

To avoid climate change, countries worldwide are trying to slow down carbon emissions and are passing stringent regulations to pressure companies to curb carbon emissions. Under pressure, companies are using carbon reduction technology to lessen carbon emissions. Reducing carbon emissions is crucial to maintain a green supply chain (Mishra et al., 2021). Many researchers have mentioned different operations that are a source of carbon emissions and methods to slow down carbon emissions. One of the first models was given by Hua et al. (2011), where they identified ordering and storage operations as a source of carbon emissions. Gurtu et al. (2015) mentioned that as the distance between buyer and vendor increases, vehicles emit more carbon and increase chances of a high carbon tax. Next to it, Tang et al. (2018) discussed methods to limit carbon emissions during transportation and manage inventory. According to Tiwari et al. (2018), carbon emissions occur due to storage, transportation, and warehousing operations in inventory management. In the same line, Sepehri et al. (2021) considered ordering and holding inventory as a source of carbon emissions in their model. Furthermore, An et al. (2021) discussed green credit financing, where a bank offers a loan to a company that invests in carbon reduction technology.

TABLE 6.1
Summary of Literature Review Related to Our Study

Paper	Demand function	Deterioration rate	No. of warehouses	Disruption	Carbon emissions
Jaggi et al. (2015)	Price dependent	Constant	Two	No	No
Sepehri et al. (2021)	Price dependent	Variable	Single	No	Yes
Aliabadi et al. (2019)	Carbon emission dependent demand	Constant		No	Yes
Rout et al. (2020)	Constant	Constant	One	No	Yes
Hovelaque and Bironneau (2015)	Price and carbon emission dependent demand	Not applicable	One	No	Yes
Mashud et al. (2021)	Price dependent	Constant and investment dependent	Single	No	Yes
Tiwari et al. (2020)	Constant	Constant	One	No	No
Gupta et al. (2020)	Constant	Variable	Two	No	No
Mishra et al. (2020)	Constant	Constant	One	No	Yes
Rana et al. (2021)	Linear time-dependent	Variable	Two	Yes	No
Present chapter	Linear time-dependent	Constant	Two	Yes	Yes

6.2.3 Deterioration Rate

Customers are more health conscious nowadays. They pay more attention to the freshness of the product. Supplying fresh products to the customer is a big challenge. The products deteriorate right from manufacturing to warehouse storage, so storing deteriorating items is again a significant challenge. Reasons for deterioration include spoilage, evaporation, damage, and expiration. Ghare and Schrader (1963), first considered deterioration rate in their EOQ (economic order quantity) model. However, Raafat et al. (1991) presented a production inventory model for deteriorating items. Further, many researchers have done excellent work in this area (Yang et al. 2010; Liang & Zhou 2011; Lee & Dye 2012; Sarkar et al. 2013). Many researchers have considered stochastic deterioration rate, which is summarized as follows (Singh & Pattnayak 2013; Bhunia & Shaikh 2014; Chakraborty et al. 2018; Rana, Kumar, Mor, et al. 2021).

Through the papers reviewed in Table 6.1, it is found that no one has considered demand disruption along with carbon emissions for deteriorating items in a two-warehouse environment. This chapter includes all these realistic phenomena and suggests investing in optimum carbon reduction technology and storing items in a rented warehouse.

6.3 PROBLEM FORMULATION

Due to the disruption in the COVID-19 situation, the entire supply chain was halted. Due to that, enormous lots of inventory were wasted. Therefore, this problem is a long-term persistent problem. We have noticed new strain in the country and COVID-19 getting worse day by day, and there is a shortage of essential items. Therefore, the situation of complete lockdown may appear shortly in the country. Hence, robust inventory models for perishable items decide order quantity and cycle duration, and to make the supply chain sustainable, carbon emissions taxation has to be considered. Therefore, we have considered two warehouses, owned warehouse (OW) and rented warehouse (RW) in perishable inventory models. The two scenarios have been considered with different duration of lockdown periods.

Governments around the globe put immense pressure on the industrial sector to reduce carbon emissions. A carbon emissions tax may affect the supply chain's overall profit. Therefore, we have sincerely considered the effect of carbon taxation on the inventory system. Carbon emissions are a result of the ordering, storage, and ambiance control of goods or items. Therefore, we have considered carbon tax to achieve minimum cost while fulfilling the demand in a disruptive scenario. Reasons for opting for a last-in-first-out (LIFO) model are high holding cost and deterioration rate in a rented warehouse compared to the owned warehouse.

6.3.1 Notations

PARAMETERS

I(t): Level of inventory at any time (t)
Z: Quantity ordered by the retailer
V: Storage capacity of the owned warehouse
T: Cycle time
T_L: Time when lockdown is imposed
T_O: Time when lockdown is lifted
T_W: Time at which the RW becomes entirely vacant
L: Lockdown period = $T_O - T_L$
Δd : Decrease in demand rate due to implementation of lockdown
Δd_1 : Upsurge in demand rate due to customer's panic buying behaviour when rules of lockdown are eased
C: Purchase cost per unit
F: Holding cost per unit per unit time in RW
H: Holding cost per unit per unit time in OW
I_{o1} (t), I_{o2} (t), I_{03} (t): Inventory level at any time t in OW in different time intervals for scenario 1 and scenario 2
I_{r1} (t), I_{r2} (t): Inventory level at any time t in different time intervals for scenario 1 and scenario 2
F: Holding cost per unit time in RW
H: Holding cost per unit time in OW
C: Purchasing cost per unit

EMISSIONS VARIABLES

B: Carbon emissions cap (kg/year)
EOC: Carbon emissions (kg/year)

σ : Reduction fraction of carbon emissions after carbon emissions investment
m: Maximum lifetime (expiry date) of items
η : Carbon tax ($/kg)
Ψ: Efficiency of carbon reduction technology
C_O : Carbon emissions unit associated in ordering (kg/year)
C_H : Carbon emissions from inventory holding in the warehouse (kg/year)

DECISION VARIABLES

G_C : Amount of capital investment in green technology ($/year)
N: (Z – V) Items stored in a rented warehouse

6.3.2 Assumptions

1. Deterioration rates in both OW and RW are constant.
2. Rate of replenishment is infinite.
3. The space of OW is only adequate to store V units, and RW can hold infinite units.
4. Firstly, OW is used to store the items to its total capacity, and then RW is used to store the excess items; that is, the LIFO dispatching policy is considered.
5. Items are first supplied from the rented warehouse when it is wholly evacuated; OW is used to satisfy the demand.
6. Items once deteriorated cannot be replaced.
7. Inventory system includes only one type of item.
8. T_L and T_O are assumed to be fixed points in time.
9. Δd and Δd_1 is assumed as a constant and known.
10. D(t): demand rate = a + bt assumed as a linear function of time, where a is the initial demand rate and b is the time-sensitivity parameter of demand rate.
11. As per Hua et al. (2011), carbon emissions occur due to warehouse ordering and holding operations.
12. Fraction of average emission reduction defined as X = $\sigma\left(1-e^{-mG_C}\right)$, X = 0, when $G_C = 0$, and X = σ $when\ G_C\Delta\infty$ carbon reduction function X (G_C) is continuously differentiable with $X\left(G_C\right)'$ >0 and $X\left(G_C\right)''$ <0.

6.4 MATHEMATICAL MODELING AND ANALYSIS

4.1 Scenario 1: Quantity ordered by the retailers is Z units, which exceeds the owned warehouse (OW) capacity (V units) of the retailer, so the remaining (Z – V) extra units are stored in RW. In this study, inventory from RW are dispatched first to fulfill customer demand. When RW becomes entirely vacant, items are dispatched from OW. At T_L, complete lockdown is imposed, and at T_O, lockdown rules are

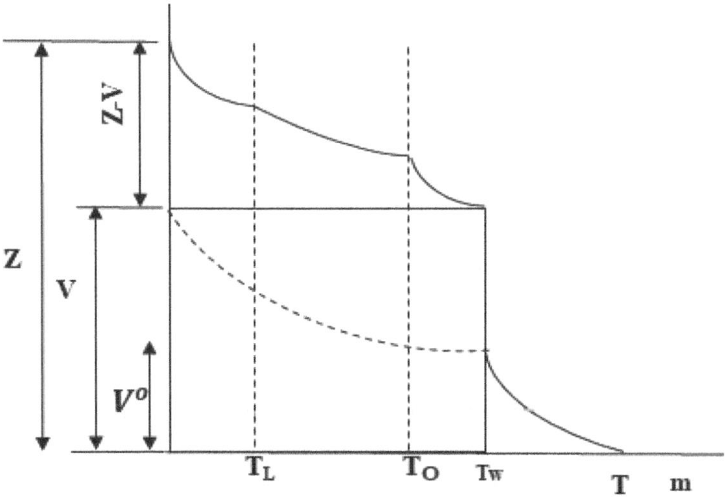

FIGURE 6.1 Graphical representation of inventory depletion for scenario 1.

eased. Both T_L and T_O lie between 0 and T_W as the items are being dispatched from RW.

For the interval 0 to T_L:

In rented warehouses, demand is smooth, and inventory depletes due to demand and deterioration. Governing differential equation is given by equation (6.1):

$$\frac{d}{dt}I_{r1}(t) = -\left(\alpha\, I_{r1}(t) + a + bt\right) \tag{6.1}$$

Applying boundary conditions $I_{r1}(0) = Z - V$:

$$I_{r1}(t) = (Z - V)e^{-\alpha t} + \left(\frac{b}{\alpha^2} - \frac{a}{\alpha}\right)\left(1 - e^{-\alpha t}\right) - \frac{b}{\alpha}t \tag{6.2}$$

Inventory left at T_L in rented warehouse:

$$I_{r1}(T_L) = (Z - V)e^{-\alpha T_L} + \left(\frac{b}{\alpha^2} - \frac{a}{\alpha}\right)\left(1 - e^{-\alpha T_L}\right) - \frac{b}{\alpha}T_L \tag{6.3}$$

But in owned warehouse items deplete due to deterioration only; governing differential equation is given by equation (6.4):

$$\frac{d}{dt}I_{o1}(t) = -\beta I_{o1}(t) \tag{6.4}$$

Applying boundary conditions for OW, $I_o(0) = V$:

$$I_{o1}(t) = Ve^{-\beta t} \tag{6.5}$$

$$I_{o1}(T_L) = Ve^{-\beta T_L} \tag{6.6}$$

For the interval T_L to T_O:

Items are being supplied from rented warehouse to fulfill the demand, and a sudden fall in order is observed due to lockdown; inventory change is given by differential equation (6.7):

$$\frac{d}{dt}I_{r2}(t) = -(\alpha I_{r2}(t) + a + bt - \Delta d) \tag{6.7}$$

Using continuity at T_L, $I_{r1}(T_L) = I_{r2}(T_L)$:

$$I_{r2}(t) = (Z - V)e^{-\alpha t} + \left(\frac{b}{\alpha^2} - \frac{a}{\alpha}\right)\left(1 - e^{-\alpha t}\right) - \frac{b}{\alpha}t + \frac{\Delta d}{\alpha}\left(1 - e^{-\alpha(t - T_L)}\right) \tag{6.8}$$

Items left at T_O in rented warehouse:

$$I_{r2}(T_O) = (Z - V)e^{-\alpha T_O} + \left(\frac{b}{\alpha^2} - \frac{a}{\alpha}\right)\left(1 - e^{-\alpha T_O}\right) - \frac{b}{\alpha}T_O + \frac{\Delta d}{\alpha}\left(1 - e^{-\alpha(T_O - T_L)}\right) \tag{6.9}$$

For owned warehouse, governing differential equation and its solution is the same as for the interval 0 to T_L:

$$\frac{d}{dt}I_o(t) = -\beta I_o(t) \tag{6.10}$$

$$I_{O1}(t) = Ve^{-\beta t} \tag{6.11}$$

Inventory left in the owned warehouse at T_O :

$$I_{o1}(T_O) = Ve^{-\beta T_O} \tag{6.12}$$

In the interval T_O to T_w, items from rented warehouses are being supplied to satisfy demand. Demand rate increases by magnitude Δd_1 as the lockdown is lifted. Inventory depletion is given by equation (6.13):

$$\frac{d}{dt}I_{r3}(t) = -(\alpha I_{r3}(t) + a + bt + \Delta d_1) \tag{6.13}$$

Applying continuity at T_O, the solution obtained is given by:

$$I_{r3}(t)=\frac{-(a+\Delta d_1)}{\alpha}-b\left(\frac{t}{\alpha}-\frac{1}{\alpha^2}\right)+\left\{\left(Z-V+\frac{a}{\alpha}-\frac{b}{\alpha^2}\right)\right.$$
$$\left.-\frac{\Delta d}{\alpha}e^{\alpha T_L}+\frac{(\Delta d+\Delta d_1)}{\alpha}e^{\alpha T_O}\right\}e^{-\alpha t} \tag{6.14}$$

Inventory level $I_{r3}(t)$ at T_W :

$$I_{r3}(T_W)=\frac{-(a+\Delta d_1)}{\alpha}-b\left(\frac{T_W}{\alpha}-\frac{1}{\alpha^2}\right)+\left\{\left(Z-V+\frac{a}{\alpha}-\frac{b}{\alpha^2}\right)\right.$$
$$\left.-\frac{\Delta d}{\alpha}e^{\alpha T_L}+\frac{(\Delta d+\Delta d_1)}{\alpha}e^{\alpha T_O}\right\}e^{-\alpha T_W} \tag{6.15}$$

Finding T_W: we know that $I_{r3}(T_W) = 0$, so from equation (6.15) we get:

$$T_W=\frac{\frac{b}{\alpha^2}-\frac{(a+\Delta d_1)}{\alpha}+\left(Z-V+\frac{a}{\alpha}-\frac{b}{\alpha^2}\right)-\frac{\Delta d}{\alpha}e^{\alpha T_L}+\frac{(\Delta d+\Delta d_1)}{\alpha}e^{\alpha T_O}}{\frac{b}{\alpha}+\alpha\left\{\left(Z-V+\frac{a}{\alpha}-\frac{b}{\alpha^2}\right)-\frac{\Delta d}{\alpha}e^{\alpha T_L}+\frac{(\Delta d+\Delta d_1)}{\alpha}e^{\alpha T_O}\right\}} \tag{6.16}$$

In owned warehouse, inventory depletion in this interval is given by differential equation:

$$\frac{d}{dt}I_{o3}(t)=-\beta I_{o3}(t) \tag{6.17}$$

The solution of the previous differential equation is:

$$I_{o3}(t)=Ve^{-\beta t} \tag{6.18}$$

Inventory left in owned warehouse at T_O :

$$I_{o3}(T_w)=Ve^{-\beta T_w} \tag{6.19}$$

$$V^O=Ve^{-\beta T_w} \tag{6.20}$$

For interval T_W to T: Now that the demand is being fulfilled from the warehouse, inventory depletion is given by the differential equation as follows:

$$\frac{d}{dt}I_{o4}(t)=-(\beta I_{o4}(t)+a+bt+\Delta d_1) \tag{6.21}$$

Solution of the previous differential equation is calculated using continuity $I_{o4}(T_w) = I_{o3}(T_w)$:

$$I_{o4}(t) = -\frac{(a+\Delta d_1)}{\beta} - b\left(\frac{t}{\beta} - \frac{1}{\beta^2}\right) + \left\{V^O + \frac{(a+\Delta d_1)}{\beta} + b\left(\frac{T_W}{\beta} - \frac{1}{\beta^2}\right)\right\} e^{\beta T_w} e^{-\beta t} \quad (6.22)$$

$$I_{o4}(T) = -\frac{(a+\Delta d_1)}{\beta} - b\left(\frac{t}{\beta} - \frac{1}{\beta^2}\right) + \left\{V^O + \frac{(a+\Delta d_1)}{\beta} + b\left(\frac{T_W}{\beta} - \frac{1}{\beta^2}\right)\right\} e^{\beta T_w} e^{-\beta T} \quad (6.23)$$

$$I_{o4}(T) = 0$$

$$T = \frac{\left\{V^O + \frac{(a+\Delta d_1)}{\beta} + b\left(\frac{T_w}{\beta} - \frac{1}{\beta^2}\right)\right\} e^{\beta T_w} + \frac{b}{\beta^2} - \frac{(a+\Delta d_1)}{\beta}}{\frac{b}{\beta} + \alpha\left\{V^O + \frac{(a+\Delta d_1)}{\beta} + b\left(\frac{T_W}{\beta} - \frac{1}{\beta^2}\right)\right\} e^{\beta T_w}} \quad (6.24)$$

COST COMPONENT ANALYSIS

1. Inventory holding cost in RW

From 0 to T_W:

$$HCRW = F\left[\int_0^{T_W} I_r(t)\,dt\right] \quad (6.25)$$

$$HCRW = F\left[\int_0^{T_L} I_{r1}(t)\,dt + \int_{T_L}^{T_O} I_{r2}(t)\,dt + \int_{T_O}^{T_W} I_{r3}(t)\,dt\right] \quad (6.26)$$

By substituting the corresponding values and solving them, we get:

$$= F\left[\left(\frac{b}{\alpha^2} - \frac{a}{\alpha}\right)T_L - \frac{bT_L^{\,2}}{2\alpha} + \left(Z - V + \frac{a}{\alpha} - \frac{b}{\alpha^2}\right)\frac{\left(1 - e^{-\alpha T_L}\right)}{\alpha} + \left(\frac{b}{\alpha^2} - \frac{a}{\alpha} + \frac{\Delta d}{\alpha}\right)(T_O - T_L) - \frac{b}{\alpha^2}\left(\frac{T_O^{\,2} - T_L^{\,2}}{2}\right)\right.$$

$$+\left(Z-V+\frac{a}{\alpha}-\frac{b}{\alpha^2}-\frac{\Delta d}{\alpha}e^{\alpha T_L}\right)\frac{\left(e^{-\alpha T_L}-e^{-\alpha T_O}\right)}{\alpha}$$

$$+\left(\frac{b}{\alpha^2}-\frac{(a+\Delta d_1)}{\alpha}\right)(T_W-T_O)-\frac{b}{2\alpha}\left(T_W^{\,2}-T_O^{\,2}\right)$$

$$+\left\{\left(Z-V+\frac{a}{\alpha}-\frac{b}{\alpha^2}-\frac{\Delta d}{\alpha}e^{\alpha T_L}\right)\right\}$$

$$\left.+\frac{(\Delta d+\Delta d_1)}{\alpha}e^{\alpha T_O}\right\}\frac{\left(e^{-\alpha T_O}-e^{-\alpha T_W}\right)}{\alpha}\Bigg] \qquad (6.27)$$

2. Values of the inventory holding cost in OW

From 0 to T:

$$\text{HCOW}=H\left[\int_0^T I_o(t)\,dt\right] \qquad (6.28)$$

$$\text{HCOW}=H\left[\int_0^{T_W} Ve^{-\beta t}dt+\int_{T_W}^{T} I_{o4}(t)\,dt\right] \qquad (6.29)$$

By substituting the corresponding values and solving them, we get:

$$=H\left[\frac{V}{\beta}\left(1-e^{-\beta T_W}\right)+\left(\frac{b}{\beta^2}-\frac{(a+\Delta d_1)}{\beta}\right)(T-T_W)-\frac{b}{\beta}\left(\frac{T^2-T_W^{\,2}}{2}\right)\right.$$

$$\left.+\left\{V^O+\frac{(a+\Delta d_1)}{\beta}+b\left(\frac{T_W}{\beta}-\frac{1}{\beta^2}\right)\right\}e^{\beta T_W}\frac{\left(e^{-\beta T_W}-e^{-\beta T}\right)}{\beta}\right] \qquad (6.30)$$

4. Cost for the deteriorated items in RW is:

$$=C\left[\alpha\int_0^{T_W} I_r(t)\,dt\right] \qquad (6.31)$$

$$=C\alpha\left[\left[\int_0^{T_L} I_{r1}(t)\,dt+\int_{T_L}^{T_O} I_{r2}(t)\,dt+\int_{T_O}^{T_W} I_{r3}(t)\,dt\right]\right] \qquad (6.32)$$

5. Cost for the deteriorated items in OW is:

$$\text{CDOW}=H\beta\left[\int_0^{T} I_o(t)\,dt\right] \qquad (6.33)$$

$$= H\beta\left[\int_{0}^{T_W} Ve^{-\beta t}dt + \int_{T_W}^{T} I_{o4}(t)dt\right] \quad (6.34)$$

6. Purchase cost:

$$P_C = \{W + V + D(T_S - T)\} \quad (6.35)$$

CARBON TAX

Carbon cap and tax concept have been introduced to control carbon emissions into the atmosphere to protect the environment and have sustainable economic development. The formulas and cost estimation for this is given as follows Sepehri et al. (2021):

$$Amount\ of\ emission(EOC) = \frac{C_O}{T} + C_H\frac{Z}{2} \quad (6.36)$$

$$Cost\ of\ emission = \eta\left(B - EOC\left(1 - \sigma\left(1 - e^{-mG_C}\right)\right)\right) \quad (6.37)$$

4.2 Scenario 2: In this scenario, items are dispatched in the same manner as in scenario 1, but the lockdown goes on for a longer time than in scenario 1. Here, a lockdown starts at the same point, but respite is given when OW items are supplied to fulfill the demand.

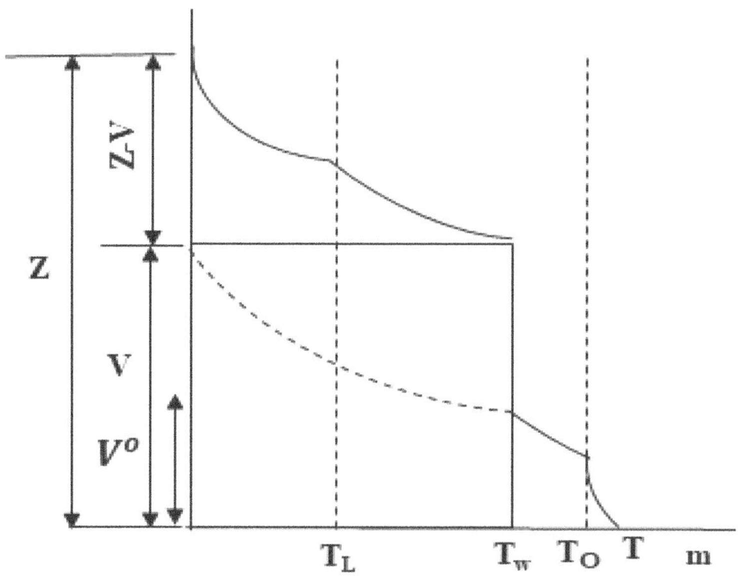

FIGURE 6.2 Graphical representation of inventory depletion for scenario 2.

For the interval 0 to T_L:

Same as in scenario 1, in rented warehouse demand is smooth and inventory depletes due to demand and deterioration. The governing differential equation is given by equation (6.38):

$$\frac{d}{dt} I_{r1}(t) = -\left(I_{r1}(t) + a + bt\right) \tag{6.38}$$

Applying boundary conditions the same as in scenario 1, $I_{r1}(0) = Z - V$:

$$I_{r1}(t) = (Z-V)e^{-\alpha t} + \left(\frac{b}{\alpha^2} - \frac{a}{\alpha}\right)\left(1 - e^{-\alpha t}\right) - \frac{b}{\alpha}t \tag{6.39}$$

Inventory left at T_L in rented warehouse:

$$I_{r1}(T_L) = (Z-V)e^{-\alpha T_L} + \left(\frac{b}{\alpha^2} - \frac{a}{\alpha}\right)\left(1 - e^{-\alpha T_L}\right) - \frac{b}{\alpha}T_L \tag{6.40}$$

For the interval, T_L to T_W demand rate is reduced by magnitude Δd same as in scenario 1.

Items are being supplied from rented warehouse to fulfill the demand, and a sudden fall in order is observed due to lockdown; same as in scenario 1, inventory change is given by differential equation (6.41)

$$\frac{d}{dt} \mathrm{I}_{r2}(\mathrm{t}) = -\left(\alpha \mathrm{I}_{r2}(\mathrm{t}) + a + b\mathrm{t} - \Delta d\right) \tag{6.41}$$

Using continuity at T_L, $I_{r1}(T_L) = I_{r2}(T_L)$:

$$\mathrm{I}_{r2}(\mathrm{t}) = (\mathrm{Z}-\mathrm{V})e^{-\alpha t} + \left(\frac{b}{\alpha^2} - \frac{a}{\alpha}\right)\left(1 - e^{-\alpha t}\right) - \frac{b}{\alpha}\mathrm{t} + \frac{\Delta d}{\alpha}\left(1 - e^{-\alpha(\mathrm{t}-\mathrm{T_L})}\right) \tag{6.42}$$

Items left in the rented warehouse at T_W are $I_{r2}(T_w)$:

$$\mathrm{I}_{r2}(\mathrm{T_w}) = 0$$

$$(\mathrm{Z}-\mathrm{V})e^{-\alpha \mathrm{T_w}} + \left(\frac{b}{\alpha^2} - \frac{a}{\alpha}\right)\left(1 - e^{-\alpha \mathrm{T_w}}\right) - \frac{b}{\alpha}\mathrm{T_w} + \frac{\Delta d}{\alpha}\left(1 - e^{-\alpha(\mathrm{T_w}-\mathrm{T_L})}\right) = 0 \tag{6.43}$$

$$\mathrm{T_w} = \frac{\left(\frac{b}{\alpha^2} - \frac{a}{\alpha}\right) + \frac{\Delta d}{\alpha} + \left(\mathrm{Z} - \mathrm{V} + \frac{a}{\alpha} - \frac{b}{\alpha^2} - \frac{\Delta d}{\alpha} e^{\alpha \mathrm{T_L}}\right)}{\frac{b}{\alpha} + \alpha\left(\mathrm{Z} - \mathrm{V} + \frac{a}{\alpha} - \frac{b}{\alpha^2} - \frac{\Delta d}{\alpha} e^{\alpha \mathrm{T_L}}\right)} \tag{6.44}$$

But in owned warehouse, items deplete due to deterioration only; the governing differential equation is given by equation (6.45):

$$\frac{d}{dt}\mathrm{I}_{o1}(t) = -\beta \mathrm{I}_{o1}(t) \tag{6.45}$$

Applying boundary conditions for OW, $\mathrm{I}_o(0) = \mathrm{V}$:

$$\mathrm{I}_{O1}(t) = Ve^{-\beta t} \tag{6.46}$$

$$\mathrm{I}_{o1}(\mathrm{T}_w) = Ve^{-\beta \mathrm{T}_w} \tag{6.47}$$

$$\mathrm{V}^{\mathrm{O}} = \mathrm{V}e^{-\beta \mathrm{T}_w} \tag{6.48}$$

For the interval $\mathrm{T_W}$ to $\mathrm{T_O}$ the decline in demand continues that started during the dispatch of items from RW:

$$\frac{d}{dt}I_{o2}(t) = -\left(\beta I_o(t) + a + bt - \Delta d\right) \tag{6.49}$$

Solution of the previous differential equation can be evaluated using continuity condition at $\mathrm{T_W}$:

$$\mathrm{I}_{o2}(\mathrm{T}_w) = \mathrm{I}_{o1}(\mathrm{T}_w)$$

$$\mathrm{I}_{o2}(t) = -\left(\frac{a-\Delta d}{\beta}\right) - b\left(\frac{t}{\beta} - \frac{1}{\beta^2}\right) + \left\{\mathrm{V}^{\mathrm{O}} + \frac{a-\Delta d}{\beta} + b\left(\frac{\mathrm{T_W}}{\beta} - \frac{1}{\beta^2}\right)\right\} e^{\beta \mathrm{T}_w} e^{-\beta t} \tag{6.50}$$

$$\mathrm{I}_{o2}(\mathrm{T}_o) = -\left(\frac{a-\Delta d}{\beta}\right) - b\left(\frac{\mathrm{T}_o}{\beta} - \frac{1}{\beta^2}\right) + \left\{\mathrm{V}^{\mathrm{O}} + \frac{a-\Delta d}{\beta} + b\left(\frac{\mathrm{T_W}}{\beta} - \frac{1}{\beta^2}\right)\right\} e^{\beta \mathrm{T}_w} e^{-\beta \mathrm{T}_o} \tag{6.51}$$

The interval $\mathrm{T_O}$ to T inventory depletion is given by equation (6.52) because the rise in demand is observed for this period due to ease of containment measures taken earlier:

$$\frac{d}{dt}\mathrm{I}_{o3}(t) = -\left(\beta \mathrm{I}_o(t) + a + bt - \Delta d_1\right) \tag{6.52}$$

Solution of the previous differential equation can be evaluated using continuity condition at T_O:

$$I_{o2}(T_O) = I_{o3}(T_O)$$

$$I_{o3}(t) = -\frac{(a+\Delta d_1)}{\beta} - b\left(\frac{t}{\beta} - \frac{1}{\beta^2}\right) + \left[\left\{V^O + \frac{a-\Delta d}{\beta} + b\left(\frac{T_w}{\beta} - \frac{1}{\beta^2}\right)\right\} e^{\beta T_w} + \frac{(\Delta d + \Delta d_1)}{\beta} e^{\beta T_O}\right] e^{-\beta t} \tag{6.53}$$

Inventory left at T in OW:

$$I_{o3}(T) = -\frac{(a+\Delta d_1)}{\beta} - b\left(\frac{T}{\beta} - \frac{1}{\beta^2}\right) + \left[\left\{V^O + \frac{a-\Delta d}{\beta} + b\left(\frac{T_w}{\beta} - \frac{1}{\beta^2}\right)\right\} e^{\beta T_w} + \frac{(\Delta d + \Delta d_1)}{\beta} e^{\beta T_O}\right] e^{-\beta T} \tag{6.54}$$

$$T = \frac{\left[\left\{V^O + \frac{a-\Delta d}{\beta} + b\left(\frac{T_w}{\beta} - \frac{1}{\beta^2}\right)\right\} e^{\beta T_w} + \frac{(\Delta d + \Delta d_1)}{\beta} e^{\beta T_O} - \frac{(a+\Delta d_1)}{\beta} + \frac{b}{\beta^2}\right]}{\frac{b}{\beta} + \beta\left[\left\{V^O + \frac{a-\Delta d}{\beta} + b\left(\frac{T_w}{\beta} - \frac{1}{\beta^2}\right)\right\} e^{\beta T_w} + \frac{(\Delta d + \Delta d_1)}{\beta} e^{\beta T_O}\right]} \tag{6.55}$$

COST COMPONENT ANALYSIS

1. Inventory holding cost in RW

From 0 to T_W:

$$HCRW = F\left[\int_0^{T_W} I_r(t)\,dt\right] \tag{6.56}$$

$$HCRW = F\left[\int_0^{T_L} I_{r1}(t)\,dt + \int_{T_L}^{T_W} I_{r2}(t)\,dt\right] \tag{6.57}$$

$$\text{HCRW} = F\left[\left(\frac{b}{\alpha^2}-\frac{a}{\alpha}\right)T_L-\frac{bT_L^{\,2}}{2\alpha}+\left(Z-V+\frac{a}{\alpha}-\frac{b}{\alpha^2}\right)\frac{\left(1-e^{-\alpha T_L}\right)}{\alpha}\right.$$

$$+\left(\frac{\Delta d}{\alpha}-\frac{a}{\alpha}+\frac{b}{\alpha^2}\right)(T_W-T_L)-\frac{b}{2\alpha}\left(T_W^{\,2}-T_L^{\,2}\right)$$

$$\left.+\left(Z-V+\frac{a}{\alpha}-\frac{b}{\alpha^2}-\frac{\Delta d}{\alpha}e^{\alpha T_L}\right)\frac{\left(e^{-\alpha T_L}-e^{-\alpha T_W}\right)}{\alpha}\right] \tag{6.58}$$

2. Inventory holding cost in OW:

$$\text{HCOW=H}\left[\int_0^{T_W} I_{O1}(t)\,dt+\int_{T_W}^{T_O} I_{O2}(t)\,dt+\int_{T_O}^{T} I_{O3}(t)\,dt\right] \tag{6.59}$$

$$HCOW = H\left[\frac{V}{\beta}\left(1-e^{-\beta T_W}\right)+\left(\frac{b}{\beta^2}-\frac{a}{\beta}+\frac{"d}{\beta}\right)(T_O-T_W)\right.$$

$$-\frac{b}{2\beta}\left(T_O^{\,2}-T_w^{\,2}\right)+\left\{V^O+\frac{a-"d}{\beta}+b\left(\frac{T_w}{\beta}-\frac{1}{\beta^2}\right)\right\}$$

$$e^{\beta T_w}\frac{\left(e^{-\beta T_W}-e^{-\beta T_O}\right)}{\beta}+\left(\frac{b}{\beta^2}-\frac{a}{\beta}-\frac{"d_1}{\beta}\right)(T-T_O)$$

$$-\frac{b}{2\beta}\left(T^2-T_O^{\,2}\right)+\left[\left\{V^O+\frac{a-"d}{\beta}+b\left(\frac{T_w}{\beta}-\frac{1}{\beta^2}\right)\right\}e^{\beta T_w}\right.$$

$$\left.\left.+\frac{("d+\Delta d_1)}{\beta}e^{\beta T_O}\right]\frac{\left(e^{-\beta T_O}-e^{-\beta T}\right)}{\beta}\right] \tag{6.60}$$

3. Deterioration cost in RW:

$$\text{DCRW=C}\left[\int_0^{T_L} I_{r1}(t)\,dt+\int_{T_L}^{T_w} I_{r2}(t)\,dt\right] \tag{6.61}$$

4. Deterioration cost in OW:

$$\text{DCOW=C}\beta\left[\int_0^{T_W} I_{O1}(t)\,dt+\int_{T_W}^{T_O} I_{O2}(t)\,dt+\int_{T_O}^{T} I_{O3}(t)\,dt\right] \tag{6.62}$$

5. Purchase cost:

$$P_C = CZ \tag{6.63}$$

CARBON TAX

A carbon cap and tax concept has been introduced to control carbon emissions into the atmosphere to protect the environment and have sustainable economic production. The formulas and cost estimation for this is given as follows Sepehri et al. (2021):

$$Amount\ of\ emission(\text{EOC}) = \frac{C_O}{\text{T}} + C_H \frac{\text{Z}}{2} \tag{6.64}$$

$$Cost\ of\ emission = \eta\left(B - EOC\left(1 - \sigma\left(1 - e^{-mG_C}\right)\right)\right) \tag{6.65}$$

8. replenishment cost per cycle = A
   TC for scenario 2

$$\text{TC}= A + EC + P_C + \text{CDOW} + \text{CDRW} + \text{HCOW} + \text{HCRW} + \text{SC} \tag{6.66}$$

6.5 SOLUTION PROCEDURE

Our objective is to derive the optimal solution for N and G_C that maximize the total average profit $TC\left(N, G_C^*\right)$.

Necessary conditions to find the optimal value for N and G_C are:

$$\frac{\partial TC(N, G_C)}{\partial N} = 0, \frac{\partial\partial TC(N, G_C)}{\partial G_C} = 0$$

Sufficient condition:

$$\frac{\partial^2 TC(N, G_C)}{\partial {G_C}^2} = J > 0, \frac{\partial^2 TC(N, G_C)}{\partial \text{N}^2} = M > 0$$

$$\frac{\partial^2 \text{TC}(G_C, \text{N})}{\partial G_C \partial \text{N}} = K = \frac{\partial^2 \text{TC}(G_C, \text{N})}{\partial \text{N} \partial G_C} = L < 0$$

$$\text{H}=\begin{bmatrix} \frac{\partial^2 \text{TC}(G_C, \text{S})}{\partial \text{N}^2} & \frac{\partial^2 \text{TC}(G_C, \text{S})}{\partial G_C \partial \text{N}} \\ \frac{\partial^2 \text{TC}(G_C, \text{S})}{\partial \text{N} \partial G_C} & \frac{\partial^2 \text{TC}(G_C, \text{S})}{\partial {G_C}^2} \end{bmatrix} = \begin{bmatrix} J & K \\ L & M \end{bmatrix}$$

For any given N and G_C if J, M > 0, and JM – K^2 > 0 then $\text{TC}(G_C, \text{S})$ is strictly convex in N and G_C and hence there exists a unique solution. But due to the highly complex equation, it isn't easy to prove this condition. Therefore, the following algorithm is used to prove convexity.

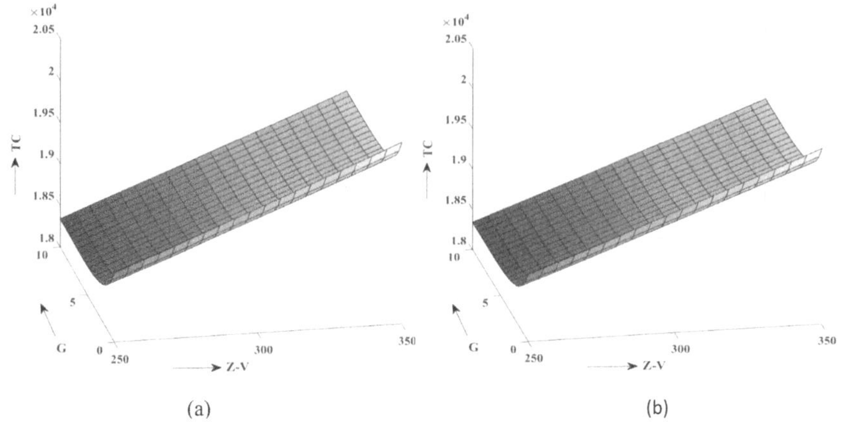

FIGURE 6.3 (a) Graphical illustration of convexity of TC w.r.t. G and Z – V for scenario 1; (b) graphical illustration of convexity of TC w.r.t. G and Z – V for scenario 2.

ALGORITHM:

Step 1: Enter all the parameters
Step 2: Initialize decision variables as [k] and [j] vector
Step 3: Perform loop operation from Step 4 to Step 9 for all discrete values of 'k' and 'j'
Step 4: Evaluate X
Step 5: Evaluate X^O
Step 6: Evaluate D(t)
Step 7: Evaluate T*w*
Step 8: Evaluate T
Step 9: Compute cost components including carbon emissions tax
Step 9: Evaluate TC
Step 10: Plot TC w.r.t. decision variables

6.6 NUMERICAL EXAMPLE

The following data sets are considered, and numerical examples are solved from the algorithm mentioned earlier.

Example 1: Data sets for scenario 1, Z = 600, V = 250, N = Z – V = 350, Δd = 150, Δd_1 = 50, a = 450, b = 25, F = 2.5, H = 2, C = 25, A = 200, T = 0.7918, T_L = 0.1, L = 0.1, $T_O = T_L + L$ = 0.2, T_W = 0.3533, S = 35, B = 10, η = 5, α = 0.09, β = 0.05, G_C = 5, n = 0.95, C_H = 1, C_O = 1, σ = 0.3
For a given value of G_C and N, TC (total cost per cycle) = $17,132.04
Example 2: Data sets for scenario 2, Z = 600, V = 250, N = Z – V = 350, Δd = 150, Δd_1 = 50, a = 450, b = 25, F = 2.5, H = 2, C = 25, A = 200,

$T = 1.2857$, $T_L = 0.1$, $L = 0.45$, $T_O = T_L + L = 0.55$, $T_W = 1.0145$, $S = 35$, $B = 10$, $\eta = 5$, $\alpha = 0.09$, $\beta = 0.05$, $G_C = 5$, $n = 0.95$, $C_H = 1$, $C_O = 1$, $\sigma = 0.3$

For a given value of G_C and N, TC (total cost per cycle) = \$17,820.02.04

For the aforementioned numerical examples, influence of the disruption period can be seen; as the disruption period increases, T and T_W increase, which leads to high storage, as well as deterioration cost in both the warehouses, so total cost per cycle TC is higher in the case of scenario 2.

Example 3: Data sets for scenario 1, V = 250, N = Z − V $\Delta d = 150$, $\Delta d_1 = 50$, a = 450, b = 25, F = 2.5, H = 2, C = 25, A = 200, T = 0.7918, $T_L = 0.1$, L = 0.1, $T_O = T_L + L = 0.2$, $T_W = 0.3533$, S = 35, B = 10, $\eta = 5$, $\alpha = 0.09$, $\beta = 0.05$, n = 0.95, $C_H = 1$, $C_O = 1$, $\sigma = 0.3$, Z = N+V

For a given value of G_C and N, TC (total cost per cycle) = \$16,751.92, $N^* - 350$, $G_C^* - 6.3158$

Example 4: Data sets for scenario 1, V = 250, $\Delta d = 150$, $\Delta d_1 = 50$, a = 450, b = 25, F = 2.5, H = 2, C = 25, A = 200, T = 0.7918, $T_L = 0.1$, L = 0.45, $T_O = T_L + L = 0.2$, $T_W = 0.3533$, S = 35, B = 10, $\eta = 5$, $\alpha = 0.09$, $\beta = 0.05$, n = 0.95, $C_H = 1$, $C_O = 1$, $\sigma = 0.3$, Z = N+V

For a given value of G_C and N, TC (total cost per cycle) = \$17,372.9, $N^* = 350$, $G_C^* = 7.8947$

With the increase in the application of carbon reduction technology, total investment costs are reduced. Still, investing in carbon reduction technology becomes costly after a specific limit, so optimum investment should be made to minimize total cost. In contrast, Tc increases with N so a small quantity should be stored in a rented warehouse.

6.7 SENSITIVITY ANALYSIS

Sensitivity analysis of both scenarios are presented in this section for behavioural analysis of TC for different parameters, such as a, Δd, $\Delta d1$, H, F, L, N = Z − V, α, β, n, N, G_C, and variation of ±20% in the value of these parameters are considered as given in Table 6.2. Data sets of numerical example 1 for scenario 1 and example 2 for scenario 2 are considered.

TABLE 6.2
The Result of Sensitivity Analysis for Both Scenarios

Parameter	Values	Scenario 1		Scenario 2	
		TC	% ΔTC	TC	% ΔTC
α	0.108	17,222.93	0.53	17,881.24	0.343
	0.072	17,029.35	0.59	17,757.11	0.353
β	0.06	17,153.24	0.12	17,874.99	0.308
	0.04	17,073.83	−0.33	17,764.08	−0.3131

(Continued)

TABLE 6.2 Continued

Parameter	Values	Scenario 1		Scenario 2	
		TC	% ΔTC	TC	% ΔTC
Δd	180	17,137.97	0.034	17,882.58	0.35
	120	17,126.15	–0.030	17,761.61	–0.32
Δd_1	60	17,074.27	–0.337	17,811.45	–0.048
	40	17,199.19	0.391	17,828.90	0.049
a	540	17,027.73	–0.608	17,560.89	–1.45
	360	17,230.52	0.574	18,194.63	2.10
F	3	17,180.13	0.280	17,905.74	0.48
	2	17,083.96	–0.280	17,734.29	–0.481
H	2.4	17,189.51	0.335	17,918.86	0.55
	1.6	17,074.58	–0.335	17,721.68	–0.551
n	1.14	17,129.04	–0.017	17,817.62	–0.013
	0.76	17,138.24	0.036	17,826.21	0.034
N	420	19,240.11	12.30	18,220.90	2.24
	280	15,053.56	–0.121	17,461.98	–2.00
G_C	6	17,129.65	- 0.01	17,817.62	–0.013
	4	17,138.24	0.036	17,826.21	0.034
L	0.54	17,138.85	0.036	17,866.92	0.263
	0.36	17,124.92	–0.04	17,782.25	–0.211

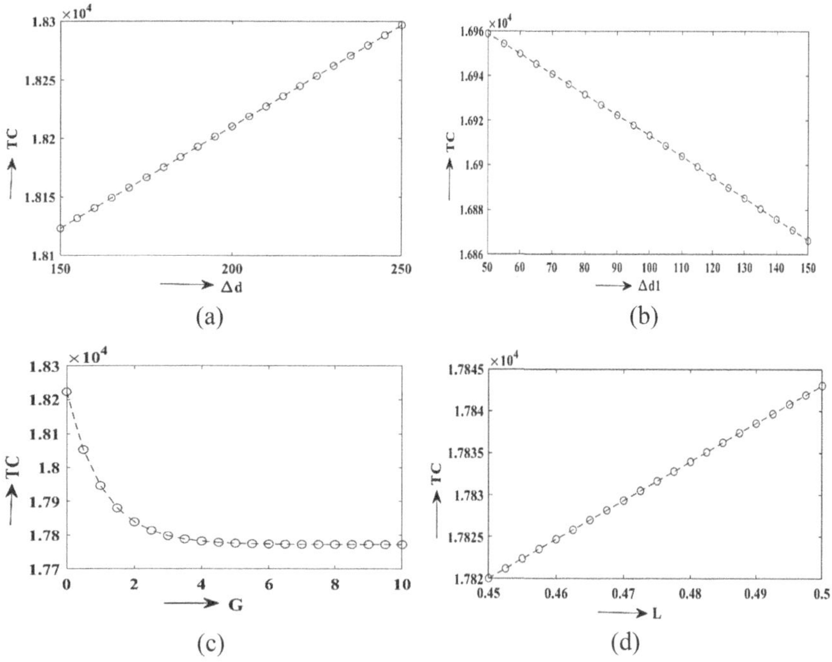

FIGURE 6.4 Variation of TC in Scenario 1 w.r.t. different parameters (a) TC and Δd, (b) TC and $\Delta d1$, (c) TC and G, and (d) TC and L.

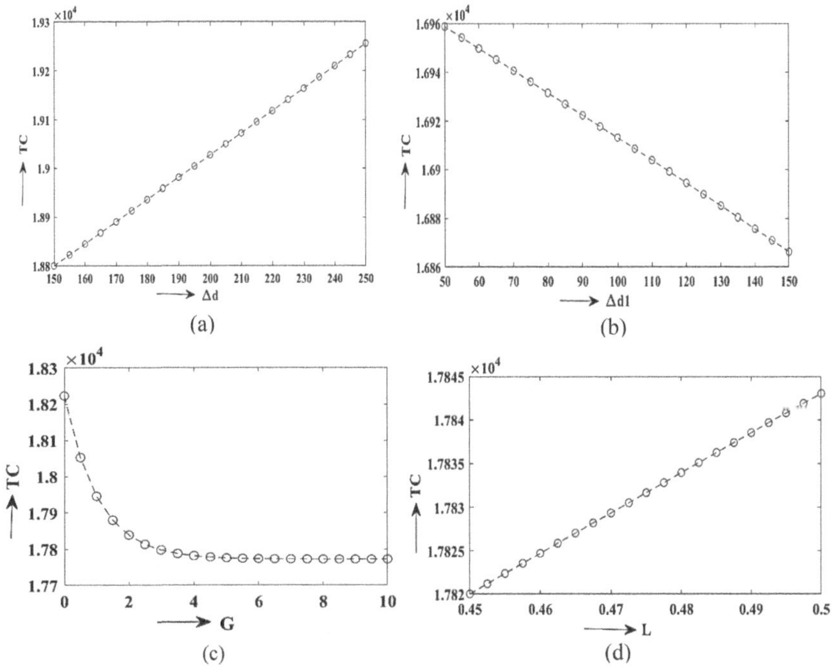

FIGURE 6.5 Variation of TC in Scenario 2 w.r.t. different parameters (a) TC and Δd, (b) TC and $\Delta d1$, (c) TC and G, and (d) TC and L.

- From the computational result obtained from sensitivity analysis, it can be seen that total cost per cycle (TC) increases and decreases with increment and decrement in deterioration rate in RW (α) and deterioration rate in OW (β), respectively.
- The effect of α is more significant as compared to β because the value of α is greater than β.
- A slight increase in the magnitude of demand disruption (Δd) results in a significant upsurge in TC. Its influence is more detrimental in scenario 2 because the duration of disruption is more extensive in scenario 2.
- Increment in Δd_1 reduces TC because the increase in demand by Δd_1 when containment measures are eased helps sell the items locked in the warehouses during the lockdown, which reduces holding and deterioration costs.
- Increment in demand rate (a) reduces the TC, whereas increment in holding cost per unit item in RW (F) and holding cost per unit item in OW (H) increases TC.
- As the number of items stored in RW (N) is increasing, TC is drastically growing so it is advised to order a small quantity (Z) so that a maximum number of items should be stored in OW, and a very small amount should be stored in RW.
- The lockdown period (L) affects the TC immensely; TC increases enormously with L.

- TC decreases with an increase in carbon reduction technology investment (G_C) and becomes asymptotic to G_C, and after a specific limit spending more on carbon reduction technology will no longer be beneficial, so a trade-off should be done whereas TC decreases with an expiry date of the product (n).

6.8 MANAGERIAL IMPLICATIONS

Strategically this model can be implemented in any inventory system to evaluate the losses caused by containment measures imposed to curb the spread of COVID-19. In case of disruption, what should be the amount of excess quantity than the warehouse's storage capacity owned by the retailer? This study can help the retailer to decide optimum investment in carbon reduction technology investment.

6.9 CONCLUSION AND FUTURE SCOPE

This chapter presents a mathematical model with a total cost per cycle to determine optimum investment in carbon reduction technology and no items to be stored in a rented warehouse. In this model, the influence of disrupted demand, both drop and upsurge in demand rate due to containment measures taken to control the propagation of COVID-19, outcome of strict lockdown duration on total cost, and a time-dependent demand in two warehouse environments are included. Two scenarios are formed to study the impact of demand disruption and lockdown period. Numerical examples are solved to validate the model. We found that in scenario 2 the total cost is higher than in scenario 1. This is because in scenario 1 the period of lockdown is small, leading to small holding and deterioration costs.

This chapter has some limitations, such as considering deterministic demand and known fluctuation in demand that in actual conditions are indefinite; deterioration rates are constant; demand function is an only linear function of time.

Based on the aforementioned limitations, there is ample future scope for this model. Demand function and deterioration rate can be stochastic; demand function can be price dependent, stock dependent, price and stock dependent, and price and freshness dependent. Inflation and trade credit can also be incorporated in this model.

REFERENCES

Aliabadi, L., Yazdanparast, R., & Nasiri, M. M. (2019). An inventory model for non-instantaneous deteriorating items with credit period and carbon emission sensitive demand: A signomial geometric programming approach. *International Journal of Management Science and Engineering Management*, *14*(2), 124–136. https://doi.org/10.1080/17509653.2018.1504331

Al-Mansour, J. F., & Al-Ajmi, S. A. (2020). Coronavirus'COVID-19'-Supply chain disruption and implications for strategy, economy, and management. *Journal of Asian Finance, Economics and Business*, *7*(9), 659–672. https://doi.org/10.13106/JAFEB.2020.VOL7.NO9.659

Al-Omoush, K. S., Simón-Moya, V., & Sendra-García, J. (2020). The impact of social capital and collaborative knowledge creation on e-business proactiveness and organizational agility in responding to the COVID-19 crisis. *Journal of Innovation & Knowledge*, *5*(4), 279–288. https://doi.org/10.1016/j.jik.2020.10.002

An, S., Li, B., Song, D., & Chen, X. (2021). Green credit financing versus trade credit financing in a supply chain with carbon emission limits. *European Journal of Operational Research*, *292*(1), 125–142. https://doi.org/10.1016/j.ejor.2020.10.025

Bhunia, A. K., & Shaikh, A. A. (2014). A deterministic inventory model for deteriorating items with selling price dependent demand and three-parameter weibull distributed deterioration. *International Journal of Industrial Engineering Computations*, *5*(3), 497–510. https://doi.org/10.5267/j.ijiec.2014.2.002

Birkie, S. E., & Trucco, P. (2020). Do not expect others do what you should! Supply chain complexity and mitigation of the ripple effect of disruptions. *International Journal of Logistics Management*, *31*(1), 123–144. https://doi.org/10.1108/IJLM-10-2018-0273

Chakrabarty, T., Giri, B. C., & Chaudhuri, K. S. (1998). An EOQ model for items with weibull distribution deterioration, shortages and trended demand: An extension of Philip's model. *Computers and Operations Research*, *25*(7–8), 649–657. https://doi.org/10.1016/s0305-0548(97)00081-6

Chakraborty, D., Jana, D. K., & Roy, T. K. (2018). Two-warehouse partial backlogging inventory model with ramp type demand rate, three-parameter Weibull distribution deterioration under inflation and permissible delay in payments. *Computers and Industrial Engineering*, *123*, 157–179. https://doi.org/10.1016/j.cie.2018.06.022

Diabat, A., & Simchi-Levi, D. (2009). A carbon-capped supply chain network problem. *IEEM 2009 – IEEE International Conference on Industrial Engineering and Engineering Management*, 523–527. https://doi.org/10.1109/IEEM.2009.5373289

Dolgui, A., & Ivanov, D. (2021). Ripple effect and supply chain disruption management: New trends and research directions. *International Journal of Production Research*, *59*(1), 102–109. https://doi.org/10.1080/00207543.2021.1840148

Donthu, N., & Gustafsson, A. (2020). Effects of COVID-19 on business and research. *Journal of Business Research*, *117*, 284–289. https://doi.org/10.1016/j.jbusres.2020.06.008

Ghare, P. M., & Schrader, G. F. (1963). A model for exponentially decaying inventory. *Journal of Industrial Engineering*, *14*, 238–243. https://ci.nii.ac.jp/naid/10004591187/en/

Grida, M., Mohamed, R., & Zaied, A. N. H. (2020). Evaluate the impact of COVID-19 prevention policies on supply chain aspects under uncertainty. *Transportation Research Interdisciplinary Perspectives*, *8*. https://doi.org/10.1016/j.trip.2020.100240

Gupta, M., Tiwari, S., & Jaggi, C. K. (2020). Retailer's ordering policies for time-varying deteriorating items with partial backlogging and permissible delay in payments in a two-warehouse environment. *Annals of Operations Research*, *295*(1), 139–161. https://doi.org/10.1007/s10479-020-03673-x

Gurtu, A., Jaber, M. Y., & Searcy, C. (2015). Impact of fuel price and emissions on inventory policies. *Applied Mathematical Modelling*, *39*(3), 1202–1216. https://doi.org/10.1016/j.apm.2014.08.001

Hovelaque, V., & Bironneau, L. (2015). The carbon-constrained EOQ model with carbon emission dependent demand. *International Journal of Production Economics*, *164*, 285–291. https://doi.org/10.1016/j.ijpe.2014.11.022

Hua, G., Cheng, T. C. E., & Wang, S. (2011). Managing carbon footprints in inventory management. *International Journal of Production Economics*, *132*(2), 178–185. https://doi.org/10.1016/j.ijpe.2011.03.024

Jaggi, C. K., Pareek, S., Khanna, A., & Sharma, R. (2015). Two-warehouse inventory model for deteriorating items with price-sensitive demand and partially backlogged shortages under inflationary conditions. *International Journal of Industrial Engineering Computations*, *6*(1), 59–80. https://doi.org/10.5267/j.ijiec.2014.9.001

Koch, J., Frommeyer, B., & Schewe, G. (2020). Online shopping motives during the COVID-19 pandemic – lessons from the crisis. *Sustainability (Switzerland)*, *12*(24), 1–20. https://doi.org/10.3390/su122410247

Lee, Y.-P., & Dye, C.-Y. (2012). An inventory model for deteriorating items under stock-dependent demand and controllable deterioration rate. *Computers & Industrial Engineering*, *63*(2), 474–482. https://doi.org/10.1016/j.cie.2012.04.006

Li, Y., Chen, K., Collignon, S., & Ivanov, D. (2021). Ripple effect in the supply chain network: Forward and backward disruption propagation, network health and firm vulnerability. *European Journal of Operational Research*, *291*(3), 1117–1131. https://doi.org/10.1016/j.ejor.2020.09.053

Li, Y., & Zobel, C. W. (2020). Exploring supply chain network resilience in the presence of the ripple effect. *International Journal of Production Economics*, *228*, 107693. https://doi.org/10.1016/j.ijpe.2020.107693

Liang, Y., & Zhou, F. (2011). A two-warehouse inventory model for deteriorating items under conditionally permissible delay in payment. *Applied Mathematical Modelling*, *35*(5), 2221–2231. https://doi.org/10.1016/j.apm.2010.11.014

Mahajan, K., & Tomar, S. (2021). COVID-19 and supply chain disruption: Evidence from food markets in India†. *American Journal of Agricultural Economics*, *103*(1), 35–52. https://doi.org/10.1111/ajae.12158

Mashud, A. H. M., Roy, D., Daryanto, Y., Chakrabortty, R. K., & Tseng, M.-L. (2021). A sustainable inventory model with controllable carbon emissions, deterioration and advance payments. *Journal of Cleaner Production*, *296*, 126608. https://doi.org/10.1016/j.jclepro.2021.126608

Messina, D., Barros, A. C., Soares, A. L., & Matopoulos, A. (2020). An information management approach for supply chain disruption recovery. *The International Journal of Logistics Management*, *31*(3), 489–519. https://doi.org/10.1108/IJLM-11-2018-0294

Mishra, U., Wu, J.-Z., & Sarkar, B. (2020). A sustainable production-inventory model for a controllable carbon emissions rate under shortages. *Journal of Cleaner Production*, *256*, 120268. https://doi.org/10.1016/j.jclepro.2020.120268

Mishra, U., Wu, J.-Z., & Sarkar, B. (2021). Optimum sustainable inventory management with backorder and deterioration under controllable carbon emissions. *Journal of Cleaner Production*, *279*, 123699. https://doi.org/10.1016/j.jclepro.2020.123699

Mondal, B., Bhunia, A. K., & Maiti, M. (2003). An inventory system of ameliorating items for price dependent demand rate. *Computers & Industrial Engineering*, *45*(3), 443–456. https://doi.org/10.1016/S0360-8352(03)00030-5

Mor, R. S., Srivastava, P. P., Jain, R., Varshney, S., & Goyal, V. (2020). Managing food supply chains post COVID-19: A perspective. *International Journal of Supply and Operations Management*, *7*(3), 295–298. https://doi.org/10.22034/IJSOM.2020.3.7

Qian, J. (2020). Ecommerce trends during Covid-19. *13*(2), 1–25. https://www.contactpigeon.com/cp/resources/ebooks/ecommerce-trends-covid19.pdf

Raafat, F. (Fred), Wolfe, P. M., & Eldin, H. K. (1991). An inventory model for deteriorating items. *Computers & Industrial Engineering*, *20*(1), 89–94. https://doi.org/10.1016/0360-8352(91)90043-6

Rana, R. S., Kumar, D., Mor, R. S., & Prasad, K. (2021). Modelling the impact of demand disruptions on two warehouse perishable inventory policy amid COVID-19 lockdown. *International Journal of Logistics Research and Applications*, 1–24. https://doi.org/10.1080/13675567.2021.1892043

Rana, R. S., Kumar, D., & Prasad, K. (2021). Two warehouse dispatching policies for perishable items with freshness efforts, inflationary conditions and partial backlogging. *Operations Management Research*. https://doi.org/10.1007/s12063-020-00168-7

Rout, C., Paul, A., Kumar, R. S., Chakraborty, D., & Goswami, A. (2020). Cooperative sustainable supply chain for deteriorating item and imperfect production under different carbon emission regulations. *Journal of Cleaner Production*, *272*, 122170. https://doi.org/10.1016/j.jclepro.2020.122170

Sarkar, B., Saren, S., & Wee, H.-M. (2013). An inventory model with variable demand, component cost and selling price for deteriorating items. *Economic Modelling*, *30*, 306–310. https://doi.org/10.1016/j.econmod.2012.09.002

Sepehri, A., Mishra, U., Tseng, M. L., & Sarkar, B. (2021). Joint pricing and inventory model for deteriorating items with maximum lifetime and controllable carbon emissions under permissible delay in payments. *Mathematics*, *9*(5), 1–27. https://doi.org/10.3390/math9050470

Shahed, K. S., Azeem, A., Ali, S. M., & Moktadir, M. A. (2021). A supply chain disruption risk mitigation model to manage COVID-19 pandemic risk. *Environmental Science and Pollution Research*. https://doi.org/10.1007/s11356-020-12289-4

Shaikh, A. A., Ca, L. E., & Tiwari, S. (2017). A two-warehouse inventory model for non-instantaneous deteriorating items with interval-valued inventory costs and stock-dependent demand under inflationary conditions. *Neural Computing & Applications*. https://doi.org/10.1007/s00521-017-3168-4

Singh, T., & Pattnayak, H. (2013). An EOQ inventory model for deteriorating items with varying trapezoidal type demand rate and Weibull distribution deterioration. *Journal of Information and Optimization Sciences*, *34*(6), 341–360. https://doi.org/10.1080/02522667.2013.838445

Tang, S., Wang, W., Cho, S., & Yan, H. (2018). Reducing emissions in transportation and inventory management: (R, Q) Policy with considerations of carbon reduction. *European Journal of Operational Research*, *269*(1), 327–340. https://doi.org/10.1016/j.ejor.2017.10.010

Tiwari, S., Cárdenas-Barrón, L. E., Shaikh, A. A., & Goh, M. (2020). Retailer's optimal ordering policy for deteriorating items under order-size dependent trade credit and complete backlogging. *Computers & Industrial Engineering*, *139*, 105559. https://doi.org/10.1016/j.cie.2018.12.006

Tiwari, S., Daryanto, Y., & Wee, H. M. (2018). Sustainable inventory management with deteriorating and imperfect quality items considering carbon emission. *Journal of Cleaner Production*, *192*, 281–292. https://doi.org/10.1016/j.jclepro.2018.04.261

Wang, Y., Wang, J., & Wang, X. (2020). COVID-19, supply chain disruption and China's hog market: A dynamic analysis. *China Agricultural Economic Review*, *12*(3), 427–443. https://doi.org/10.1108/CAER-04-2020-0053

Yang, H.-L., Teng, J.-T., & Chern, M.-S. (2010). An inventory model under inflation for deteriorating items with stock-dependent consumption rate and partial backlogging shortages. *International Journal of Production Economics*, *123*(1), 8–19. https://doi.org/10.1016/j.ijpe.2009.06.041

7 Development of Software Prototype for Supplier Selection amid COVID-19 Pandemic

Kanika Prasad and Samidha Prasad

CONTENTS

7.1 INTRODUCTION

The current situation with the COVID-19 pandemic is dominating many spheres of our lives and businesses. The pandemic has led to the temporary closure of many manufacturing firms due to unavailability of raw materials, causing disruption in the related supply chain. The lockdown imposed by the governments to limit the spread of the pandemic has also affected the delivery of products to end users. The majority of companies are finding it difficult to maintain a steady flow of even fast-moving consuming goods. In order to make up for the shortage of raw materials, the manufacturing firms need to look for alternative suppliers. Under these circumstances, the role of a supplier in a business becomes much more demanding as high-quality products are required at the retailers' end, and producers look forward to suppliers to sell a larger quantity of products. Owing to this, the suppliers must be flexible and good at managing relationships. Suppliers form an indispensable part of any business, and their role is crucial in maximizing business performance. Generally, a manufacturer has many suppliers but most of them are affected by the COVID-19 pandemic and the persisting suppliers may not be able to deliver enough supplies to the manufacturer. Therefore, selection of an appropriate supplier in the current situation of the COVID-19 pandemic has become a key task for different organizations.

DOI: 10.1201/9781003150084-7

The selection of a suitable supplier not only enhances supply chain sustainability but also improves the quality and efficiency of the supply chain at reduced cost (Kannan et al. 2020). The globalized business process has increased the supplier base. Moreover, several operational, health-related and other factors specific to the COVID-19 pandemic need to be considered while making such decisions. These factors often have complex relationships among them. This large supplier base together with conflicting criteria often convert this decision-making process to a difficult and time-consuming task, thereby making it a multi-criteria decision-making (MCDM) problem (Rao 2007). Hence, in the process of selecting the most appropriate supplier by a manufacturer, an optimum combination of factors is usually sought, rather than one particular criterion. Thus, there is an ardent need for a well-structured systematic approach that will assist the organizations in choosing appropriate suppliers. The approach must take into account the factors that address the issues related to global disruptions, such as COVID-19. Therefore, in this work, a graphical user interface (GUI) based on the VIKOR (Vlse Kriterijumska Optimizacija Kompromisno Resenje) method is designed. The platform used for this purpose is Visual BASIC 6.0. The proposed methodology would help in the supplier selection under the global disruption conditions similar to COVID-19. Other than the traditional economic, social and environmental criteria, the present study considers additional factors which play an important role while selecting a suitable supplier in the current situation of pandemic. The developed software prototype would ease the supplier selection decision-making process under the global disruption conditions similar to COVID-19.

7.2 LITERATURE REVIEW

The current situation of the COVID-19 pandemic has caused many disturbances in the current supply chain. As a result, many suppliers have been forced to shut down, because of which the contracts signed with these suppliers stand either cancelled or breached. In this condition the buyer is left with no other option but to look for another supplier. Further, the important prerequisites for solving any selection problem are identification of appropriate criteria and suitable methods. Therefore, the literature review presented here is organized into three different segments as per the research objectives. In the first section, a review on understanding the impact of COVID-19 on the global supply chain is presented. In the next section, an overview of the selection criteria utilized in supplier selection problems is provided, and in the last section a review of the methods employed for supplier selection is presented.

Ivanov (2020) evaluated and predicted the influence of the COVID-19 pandemic on the supply chain worldwide through conducting a simulation study. Hoek (2020) considered seven companies that implemented several managerial steps to mitigate the effect of the COVID-19 pandemic. The study concluded that it is more important to look for an alternative supplier in order minimize risks in the supply chain. Sharma et al. (2020) identified six criteria which affect supply chains amid the COVID-19 pandemic. A framework based on those criteria was also developed for enhancing the sustainability of supply chains. Majumdar et al. (2020) used a case study from textile supply chains in South Asian countries to focus on the fragility of supply chain in this COVID-19 pandemic and gave a few suggestions based on the study. Ivanov and Dolgui (2020) suggested that intertwined supply chains would provide more sustainable services to society.

Identification of relevant and suitable criteria for choosing a right supplier is an important step in the whole process. No official partnership exists with alternative suppliers. Under normal conditions, the best choice will not always be an alternative supplier, but that must be taken into consideration during the COVID-19 pandemic. Hence, there would be few additional factors that affect the supplier selection (Chen and Lin 2020). In this study, along with social, economic and environmental factors, aspects critical to the COVID-19 pandemic are also taken into consideration. Performance in containing the pandemic, its severity, infection risk, vaccine procurement speed, supplier impact and demand shrinkage are some of the criteria relevant to the COVID-19 pandemic (Chen et al. 2020). The important criteria for supplier selection used in this study is presented in Table 7.1.

An assorted list of different MCDM methods that have been successfully applied for supplier selection in last few years is displayed in Table 7.2. It can be concluded after the literature review that the methods commonly used for choosing a supplier are Analytic Hierarchy Process (AHP) and Technique for Order of Preference by Similarity to Ideal Solution (TOPSIS). However, as per the best of authors' knowledge, the literature is scarce in terms of development of software prototype with a user-friendly GUI for supplier selection. Alavi et al. (2021) designed a decision support system for choosing a suitable supplier in the circular supply chain, but this system does not incorporate factors critical to the COVID-19 pandemic and similar situations. Hence, in this research work, a software prototype based on the VIKOR methodology is designed that can automate the supplier selection process in situations similar to the COVID-19 pandemic taking disruption-related, social and emotional criteria into consideration.

7.3 DESIGN OF A VIKOR-BASED SOFTWARE PROTOTYPE FOR SUPPLIER SELECTION

7.3.1 Methodology

VIKOR is a Serbian word that implies optimization and compromise solution in an MCDM environment. Opricovic (1998) introduced this method as an efficient technique aimed at choosing and prioritizing a list of alternatives based on conflicting criteria. This very popular tool is being used extensively in the field of MCDM owing to its computational simplicity and accuracy of the results obtained. It has proved to be an efficient tool in conditions when the decision maker is unable to agree upon his/her preference in the beginning phase. A specified value of nearness to the ideal solution is considered to determine the compromise ranking list. Also, the result obtained is accepted as it offers a maximum "group utility" of the "major portion", and a minimum of "discrete regret" of the "remaining". Depending upon the preference of the decision maker on criteria weight, the compromise solutions are negotiated (Chatterjee and Chakraborty 2016). Different versions of the VIKOR method have been developed to cater to the need of decision-making problems. In the current work

TABLE 7.1
Criteria for Supplier Selection

Criteria	Price	Product quality	Delivery speed	Company reputation	Technological capacity	Pollution control and recycling	Recyclable raw materials for production	Respecting environmental standards	Waste management	Human rights	Labour health and work safety	Pandemic containment performance	Severity of pandemic	Vaccine procurement speed
Fallahpour et al. (2017)		√	√			√		√	√	√	√			
Luthra et al. (2017)	√	√	√		√	√	√		√	√	√			
Awasthi et al. (2018)	√	√				√	√		√	√				
Azimifard et al. (2018)	√	√		√	√	√				√				
Kannan (2018)	√	√	√		√	√	√	√	√	√	√			
Vahidi et al. (2018)	√				√	√		√	√		√			
Ahmadi and Amin (2019)	√	√	√	√	√	√	√	√	√					
Alikhani et al. (2019)	√	√	√	√	√		√	√		√				

Guarnieri and Trojan (2019)	√	√	√		√	√			√	√				
Li et al. (2019)	√	√	√			√	√			√	√			
Memari et al. (2019)	√		√	√	√	√	√	√	√	√	√			
Mohammed et al. (2019)	√	√	√		√	√			√		√			
Pishchulov et al. (2019)		√	√		√	√			√		√			
Yu et al. (2019)	√	√	√		√	√	√			√	√			
Govindan et al. (2020)		√	√			√	√	√						
Jia et al. (2020)	√	√	√		√	√			√		√			
Kannan et al. (2020)	√	√	√	√	√	√	√	√		√	√			
Stević et al. (2020)	√	√	√	√		√	√	√	√	√	√			
Chen et al. (2021)	√	√	√	√								√	√	√

TABLE 7.2
MCDM Methods for Supplier Selection

MCDM methods	AHP/fuzzy AHP	ANP/fuzzy ANP	TOPSIS/ fuzzy TOPSIS	VIKOR	BWM/ fuzzy BWM	DEA	DEMATEL	ELECTRE	MARCOS/D-MARCOS	SWOT-QFD
Fallahpour et al. (2017)	√		√							
Luthra et al. (2017)	√			√						
Awasthi et al. (2018)	√			√						
Azimifard et al. (2018)	√		√							
Vahidi et al. (2018)										√
Alikhani et al. (2019)				√		√				
Guarnieri and Trojan (2019)	√									
Li et al. (2019)			√							
Memari et al. (2019)			√							
Mohammed et al. (2019)	√		√							
Pishchulov et al. (2019)	√									
Yu et al. (2019)			√							
Chattopadhyay et al. (2020)									√	
Govindan et al. (2020)		√					√			
Kannan et al. (2020)				√	√					
Stević et al. (2020)									√	
Chen et al. (2021)			√							
Alavi et al. (2021)					√					

a software prototype based on the VIKOR method is developed for supplier selection amid the COVID-19 pandemic.

The first step in this method is to construct a matrix displaying the alternatives' performance corresponding to the chosen criteria. This matrix is known as a decision matrix. Suppose f_{ij} denotes the performance of ith alternative corresponding to j^{th} criterion. The L_p -metric represented by equation (7.1) is used for developing a multi-criteria degree for compromise ranking. In a compromise programming method, the L_p metric is employed as a combining function (Zeleny 1982):

$$L_{p,i} = \left\{ \sum_{j=1}^{m} \left(w_j \left[\left(f_{ij}\right)_{max} - f_{ij} \right] / \left[\left(f_{ij}\right)_{max} - \left(f_{ij}\right)_{min} \right] \right)^p \right\}^{\frac{1}{p}},$$

$$1 \leq p \leq \infty;\ i = 1,\ 2, \ldots n \tag{7.1}$$

Where, m and n represent the number of criteria and number of alternatives respectively and w_j is the weight of jth criterion. In the VIKOR method, $L_{1,i}$ and $L_{\infty,i}$ are utilized to express the ranking pre-order. The steps followed in the VIKOR method are described as follows:

a Find the best, $\left(f_{ij}\right)_{max}$, and the worst, $\left(f_{ij}\right)_{min}$, values among all the criteria from the decision matrix.

b Obtain the priority weights of criteria.

c Calculate the values of S_i and R_i while employing equations (7.2) and (7.3):

$$S_i = L_{1,i} = \sum_{j=1}^{m} w_j \left[\left(f_{ij}\right)_{max} - f_{ij} \right] / \left[\left(f_{ij}\right)_{max} - \left(f_{ij}\right)_{min} \right] \tag{7.2}$$

$$R_i = L_{\infty,i} = Max \left\{ w_j \left[\left(f_{ij}\right)_{max} - f_{ij} \right] / \left[\left(f_{ij}\right)_{max} - \left(f_{ij}\right)_{min} \right] \right\},$$

$$j = 1, 2, \ldots m \tag{7.3}$$

The criteria can be of two types: beneficial or non-beneficial in nature. Maximum values are always desired for beneficial criteria, for which equation (7.2) is applicable, while minimum values are preferred for non-beneficial criteria for which the term $[\left(f_{ij}\right)_{max} - f_{ij}]$ in equation (7.2) is replaced by $[f_{ij}] - \left(f_{ij}\right)_{min}]$. Therefore, for non-beneficial criteria, equation (7.2) can be rearranged as:

$$S_i = L_{1,i} = \sum_{j=1}^{M} w_j \left[f_{ij} - \left(f_{ij}\right)_{min} \right] / \left[\left(f_{ij}\right)_{max} - \left(f_{ij}\right)_{min} \right] \tag{7.4}$$

d Compute Q_i value:

$$Q_i = v((S_i - S_{i-min}) / (S_{i-max} - S_{i-min})) + (1 - v)$$

$$((R_i - R_{i-min}) / (R_{i-max} - R_{i-min})) \tag{7.5}$$

where S_{i-max} and S_{i-min} are the highest and lowest values of S_i respectively, and R_{i-max} and R_{i-min} are the maximum and minimum values of R_i respectively. Weight of the strategy of "the major attributes" (or "the maximum group utility") is denoted by v. The ranking of alternatives is influenced by the value of v and is usually determined independently by the decision expert. The value for weight falls in between 0 and 1. Generally, the value of v as 0.5 is considered ideal. The compromise can be chosen with "consensus" ($v = 0.5$), with "voting by majority" ($v > 0.5$), or with "veto" ($v < 0.5$).

e Arrange the alternatives in ascending order as per the derived values of S, R and Q. Three ranking lists are obtained from this method.

7.3.2 Framework for VIKOR-Based Software Prototype

A user-friendly GUI in Visual BASIC 6.0 is designed to automate the decision-making process while selecting an alternative supplier. The steps followed while using this software prototype are described as follows:

Step I – On running the application, an opening window with guidelines to aid the decision maker to navigate through it properly appears as shown in Figure 7.1.

Step II – The end user then inputs the number of alternatives and criteria in the boxes provided to generate an empty decision matrix of required dimension.

Step III – The evaluation criteria are identified as either beneficial or non-beneficial after pressing "Type of criteria" button.

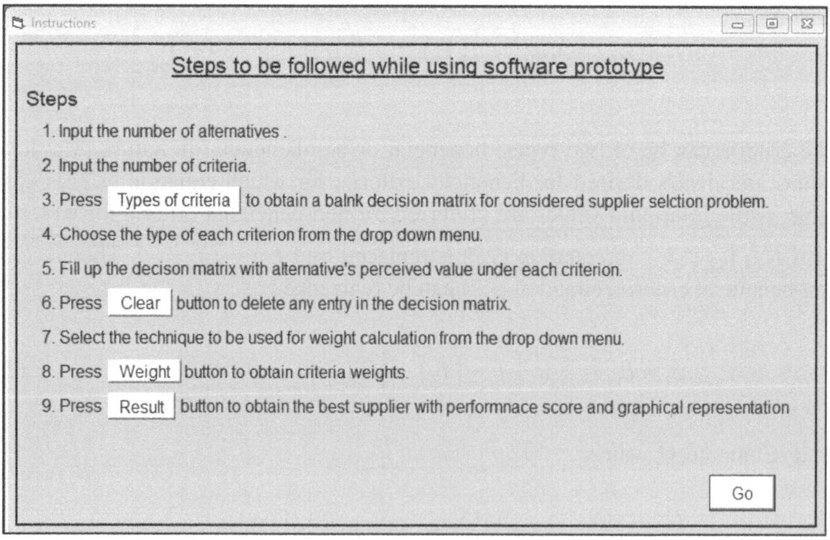

FIGURE 7.1 Opening window of the software prototype for supplier selection.

Step IV – The decision matrix is filled up with relevant details of the supplier alternatives according to the availability.
Step V – There are three different options for the end user for determining weights of evaluation criteria, that is, "Critic", Ëntropy" and "Manual". The user has the flexibility to select any of the three options.
Step VI – After choosing the weightage calculation method, the "Weight" functional key is pressed to obtain the criteria weights.
Step VII – At the end, the "Result" key is pressed to obtain the ranking of suppliers with their performance scores and graphical representation.

7.4 ILLUSTRATIVE EXAMPLE

Identifying an alternative supplier from a healthcare point of view is critical in this widespread COVID-19 pandemic. In this study, a software prototype based on the VIKOR method is developed for alternative supplier selection. In this section, an empirical study is provided from the healthcare sector to demonstrate the use of developed application for supplier selection. An example of ABC hospital based in Kahalgaon, Bihar, which provides gynaecological consultancy and care to the women, is taken into consideration. Since the materials and tools are the medical supplies that would be used for surgical procedures too, they must meet the quality standards of a healthcare board. Also, in the current times of pandemic when the infection rate is high, medical supplies with higher quality and quantity are required at the hospitals. Therefore, the selection of an alternative supplier is of great importance.

First, while solving the supplier selection problem, the suppliers capable of delivering medical supplies were determined. In the second step, out of the total identified supplier, six suppliers were shortlisted who had the potential of supplying the required amount of quality medical supplies. The process of alternative supplier selection in the medical domain is similar to other sectors of economy and, hence, criteria applicable in other industries may be easily applied. However, it is indispensable to incorporate some new criteria that will affect the selection process amid this pandemic. Table 7.3 displays a list of criteria and their definitions that have been taken

TABLE 7.3
Evaluation Criteria for Supplier Selection in Healthcare Industry

Criteria	Symbol	Definition
Price	C1	The value of product in terms of money.
Product quality	C2	Features of the product that meet customer requirements.
Delivery time	C3	Due date by which the suppliers must deliver their products.
Company reputation	C4	Stakeholder's opinion about an organization.
Respecting environmental standards	C5	Following standards that will minimize harmful impacts on the environment.
Waste management	C6	Reusing and recycling of materials and products while minimizing the waste.

(Continued)

TABLE 7.3 Continued

Criteria	Symbol	Definition
Work safety and labour health	C7	Measures taken for protecting health and life of employees.
Pandemic containment performance	C8	Shows how well the pandemic is kept in control.
Infection risk	C9	Likelihood of being infected.

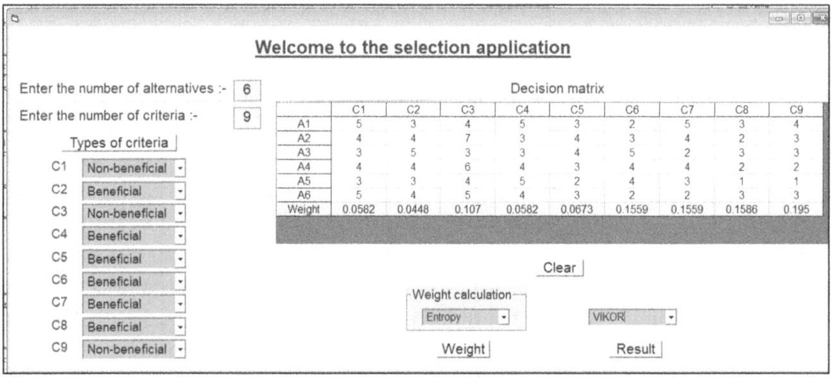

FIGURE 7.2 Decision matrix for supplier selection.

into consideration for supplier selection in this study. This decision-making problem consists of nine criteria. Three criteria – "Price", "Delivery time" and "Infection risk" – are the non-beneficial criteria for which lower values are always preferred. All the other criteria are beneficial in nature and are required to have higher values.

Subsequently, a decision matrix having the performance of potential suppliers based on the chosen criteria is developed. The information brochure provided by the suppliers and communication with other hospitals and their understanding with those suppliers were used for gathering data on suppliers. The qualitative criteria are assigned values using a 1–5 rating scale, where 1 shows worst performance and 5 denotes highest performance. The relevant decision matrix is shown in Figure 7.2.

In the next step, a priority weight for each criterion needs to be determined. In order to eliminate subjectivity in judgements, Shannon's entropy method is used for determining the priority weights. This makes the evaluation procedure unbiased. The software prototype also provides the end user with two other options – "Critic" and "Manual" – for providing the criteria weights for further calculations. The user may select any of the methods for weightage calculation from the drop-down menu depending upon the requirement. Last, upon pressing the "Result" button, an output window with performance scores and ranks of suppliers appears. Figure 7.3 displays the output window for the developed software prototype.

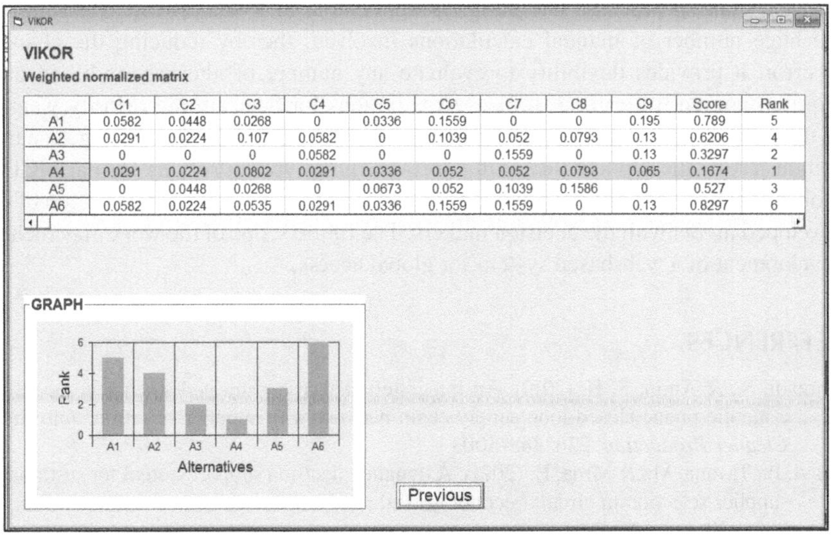

	C1	C2	C3	C4	C5	C6	C7	C8	C9	Score	Rank
A1	0.0582	0.0448	0.0268	0	0.0336	0.1559	0	0	0.195	0.789	5
A2	0.0291	0.0224	0.107	0.0582	0	0.1039	0.052	0.0793	0.13	0.6206	4
A3	0	0	0	0.0582	0	0	0.1559	0	0.13	0.3297	2
A4	0.0291	0.0224	0.0802	0.0291	0.0336	0.052	0.052	0.0793	0.065	0.1674	1
A5	0	0.0448	0.0268	0	0.0673	0.052	0.1039	0.1586	0	0.527	3
A6	0.0582	0.0224	0.0535	0.0291	0.0336	0.1559	0.1559	0	0.13	0.8297	6

FIGURE 7.3 Output window with ranking of the considered suppliers.

Finally, when these alternatives are arranged in ascending order of their performance score, a ranking list of the potential suppliers is obtained as A4>A3>A5>A2>A1>A6. Hence, it can be concluded that "Supplier 4" tops the ranking list, followed by "Supplier 3". On the other hand, "Supplier 6" must be the least preferred choice for the procurement of medical supplies. A closer look at the decision matrix reveals that "Supplier 4" surpasses the other alternatives due to its better performance on criteria such as waste management, infection risk and work safety and labour health having higher weights 0.1559, 0.195 and 0.1559 respectively, though its price and delivery times are comparatively high. On the other hand, "Supplier 6" is identified as the worst performer fibre with respect to waste management, infection risk, and work safety and labour health which are dominating criteria in the present study. The hospital under consideration faced a real-time dilemma of supplier selection amid this pandemic. After talking with the stakeholders and considering various factors that are similar to the evaluation criteria used in this study, the hospital authority arrived at the same conclusion as done by the developed software prototype. Supplier 4 was selected, which demonstrates the applicability of this developed model.

7.5 CONCLUSIONS

In the prevailing condition of the COVID-19 pandemic, it has become very important to look for alternative suppliers. However, the availability of a large number of suppliers with varying attributes makes the selection problem more complex and time consuming. Also, some of the factors which originated due to the pandemic must also be considered while selecting an alternative supplier. Therefore, in this paper, a software prototype based on the VIKOR method is developed to automate the entire decision-making process. The designed application in Visual BASIC 6.0 removes the huge number of manual calculations involved, thereby reducing the chances of error. It provides flexibility to evaluate any number of alternatives based on a required number of criteria. It also gives options for determining priority weights of the criteria. The result derived from this model corroborate the existing scenario, demonstrating the applicability of this software prototype for solving similar MCDM problems. The GUI built in Visual BASIC 6.0 facilitates a seamless interaction of the developed model with the decision makers. The future scope of the work may include development of a web-based system for global access.

REFERENCES

Ahmadi, S., & Amin, S. H. (2019). An integrated chance-constrained stochastic model for a mobile phone closed-loop supply chain network with supplier selection. *Journal of Cleaner Production*, 226, 988–1003.

Alavi, B., Tavana, M., & Mina, H. (2021). A dynamic decision support system for sustainable supplier selection in circular economy. *Sustainable Production and Consumption*, 27, 905–920.

Alikhani, R., Torabi, S. A., & Altay, N. (2019). Strategic supplier selection under sustainability and risk criteria. *International Journal of Production Economics*, 208, 69–82.

Awasthi, A., Govindan, K., & Gold, S. (2018). Multi-tier sustainable global supplier selection using a fuzzy AHP-VIKOR based approach. *International Journal of Production Economics*, 195, 106–117.

Azimifard, A., Moosavirad, S. H., & Ariafar, S. (2018). Selecting sustainable supplier countries for Iran's steel industry at three levels by using AHP and TOPSIS methods. *Resources Policy*, 57, 30–44.

Chatterjee, P., & Chakraborty, S. (2016). A comparative analysis of VIKOR method and its variants. *Decision Science Letters*, 5, 469–486.

Chattopadhyay, R., Chakraborty, S., & Chakraborty, S. (2020). An integrated D-MARCOS method for supplier selection in an iron and steel industry. *Decision Making: Applications in Management and Engineering*, 3(2), 49–69.

Chen, T., & Lin, C.-W. (2020). Smart and automation technologies for ensuring the long-term operation of a factory amid the COVID-19 pandemic: An evolving fuzzy assessment approach. *The International Journal of Advanced Manufacturing Technology*, 111, 3545–3558

Chen, T., Wang, Y. C., & Chiu, M. C. (2020). Assessing the robustness of a factory amid the COVID-19 pandemic: A fuzzy collaborative intelligence approach. *Healthcare*, 8, 481.

Chen, T., Wang, Y. -C., & Wu, H. -C. (2021). Analyzing the impact of vaccine availability on alternative supplier selection amid the COVID-19 pandemic: A cFGM-FTOPSIS-FWI approach. *Healthcare*, 9, 71.

Fallahpour, A., Olugu, E. U., Musa, S. N., Wong, K. Y., & Noori, S. (2017). A decision support model for sustainable supplier selection in sustainable supply chain management. *Computers & Industrial Engineering*, 105, 391–410.

Govindan, K., Mina, H., Esmaeili, A., & Gholami-Zanjani, S. M. (2020). An integrated hybrid approach for circular supplier selection and closed-loop supply chain network design under uncertainty. *Journal of Cleaner Production*, 242, 118317

Guarnieri, P., & Trojan, F. (2019). Decision making on supplier selection based on social, ethical, and environmental criteria: A study in the textile industry. *Resources, Conservation and Recycling*, 141, 347–361.

Hoek, R. V. (2020). Responding to COVID-19 supply chain risks-Insights from supply chain change management, total cost of ownership and supplier segmentation theory. *Logistics*, 4, 23.

Ivanov, D. (2020). Predicting the impacts of epidemic outbreaks on global supply chains: A simulation-based analysis on the coronavirus outbreak (COVID-19/SARS-CoV-2) case. *Transportation Research Part E: Logistics and Transportation Review*, 136, 101922.

Ivanov, D., & Dolgui, A. (2020). Viability of intertwined supply networks: Extending the supply chain resilience angles towards survivability. A position paper motivated by COVID-19 outbreak. *International Journal of Production Research*, 58, 2904–2915.

Jia, R., Liu, Y., & Bai, X. (2020). Sustainable supplier selection and order allocation: Distributionally robust goal programming model and tractable approximation. *Computers & Industrial Engineering*, 140, 106267

Kannan, D. (2018). Role of multiple stakeholders and the critical success factor theory for the sustainable supplier selection process. *International Journal of Production Economics*, 195, 391–418.

Kannan, D., Mina, H., Nosrati-Abarghooee, S., & Khosrojerdi, G. (2020). Sustainable circular supplier selection: A novel hybrid approach. *The Science of the Total Environment*, 722, 137936

Li, J., Fang, H., & Song, W. (2019). Sustainable supplier selection based on SSCM practices: A rough cloud TOPSIS approach. *Journal of Cleaner Production*, 222, 606–621.

Luthra, S., Govindan, K., Kannan, D., Mangla, S. K., & Garg, C. P. (2017). An integrated framework for sustainable supplier selection and evaluation in supply chains. *Journal of Cleaner Production*, 140, 1686–1698.

Majumdar, A., Shaw, M., & Sinha, S. K. (2020). COVID-19 debunks the myth of socially sustainable supply chain: A case of the clothing industry in South Asian countries. *Sustainable Production and Consumption*, 24, 150–155.

Memari, A., Dargi, A., Jokar, M. R. A., Ahmad, R., & Rahim, A. R. A. (2019). Sustainable supplier selection: A multicriteria intuitionistic fuzzy TOPSIS method. *Journal of Manufacturing Systems*, 50, 9–24.

Mohammed, A., Harris, I., & Govindan, K. (2019). A hybrid MCDM-FMOO approach for sustainable supplier selection and order allocation. *International Journal of Production Economics*, 217, 171–184.

Opricovic, S. (1998). Multicriteria optimization of civil engineering systems. *Faculty of Civil Engineering, Belgrade*, 2, 5–21.

Pishchulov, G., Trautrims, A., Chesney, T., Gold, S., & Schwab, L. (2019). The Voting Analytic Hierarchy Process revisited: A revised method with application to sustainable supplier selection. *International Journal of Production Economics*, 211, 166–179.

Rao, R. V. (2007). *Decision making in the manufacturing environment using graph theory and fuzzy multiple attribute decision making methods*. London: Springer-Verlag.

Sharma, M., Luthra, S., Joshi, S., & Kumar, A. (2020). Developing a framework for enhancing survivability of sustainable supply chains during and post-COVID-19 pandemic. *International Journal of Logistics Research and Application*, 1–21.

Stević, Ž., Pamučar, D., Puška, A., & Chatterjee, P. (2020). Sustainable supplier selection in healthcare industries using a new MCDM method: Measurement of alternatives and ranking according to compromise solution (MARCOS). *Computers & Industrial Engineering*, 140, 106231.

Vahidi, F., Torabi, S. A., & Ramezankhani, M. J. (2018). Sustainable supplier selection and order allocation under operational and disruption risks. *Journal of Cleaner Production*, 174, 1351–1365.

Yu, C., Shao, Y., Wang, K., & Zhang, L. (2019). A group decision making sustainable supplier selection approach using extended TOPSIS under interval-valued Pythagorean fuzzy environment. *Expert Systems with Applications*, 121, 1–17

Zeleny, M. (1982). *Multiple criteria decision making*. New York: McGraw Hill.

8 Improving Supply Chain Resilience under COVID-19 Outbreak through Industry 4.0

A Review on Tools and Technologies

Nikita Sinha, Mohammad Faisal Noor, and Amaresh Kumar

CONTENTS

DOI: 10.1201/9781003150084-8

8.1 INTRODUCTION

The unforeseen challenges arising in the global market due to the COVID-19 pandemic have forced business leaders to look for innovative solutions to solve the disruptions caused. This crisis has revealed the bottlenecks and vulnerabilities of an inflexible supply chain. The main challenge faced by most of the companies was being unable to accommodate any changes in demand on time (which should be the utmost priority) because of an inadequate and rigid supply chain (Ranney et al., 2020). The pandemic gave insight into the critical situations and the need to redesign the supply chain into one capable of handling disruptions in real time and operating flexibly (Sharma et al., 2020). This unprecedented situation gave rise to the urgent need for widespread integration of advanced technologies in various industries and, consequently, in the supply chain to overcome the challenges caused by any such disruptions. Novel methods are being explored to adapt to similar challenges in future, and many researchers are working in this field.

As a result of COVID-19's profound impact, a recently published article (Sherman, 2020) stated that 94% of Fortune 1000 companies experienced supply chain disruption, and many other companies suffered moderate to high setbacks. Due to inadequate facilities to meet this unforeseen surge in demand, the average lead time of operations had nearly increased twofold globally. To counter any such adversities in the future, organizations need to upgrade their standard of operations by being at par with the current developments in productivity and sustainability to remain a global player. Simultaneously, the increasing customer base has also led to the evolution of the supply chain into a globally interconnected network. Supply chain flexibility is considered one of the essential topics for research over the past few years (Burin et al., 2020). Flexibility can be comprehended as the competency of a firm to respond to long-term changes with shorter lead times at the lowest cost in the context of the supply chain by adjusting its structure under the changing environment (Eckstein et al., 2015; X. Li et al., 2009). A flexible supply chain is said to enhance a firm's operational performance while maintaining its customer satisfaction (Bertrand, 2003).

The supply chain can be streamlined by improving the flexibility attainable by integrating advanced technologies of Industry 4.0. These technologies include big data, cloud computing, augmented/virtual reality, and artificial intelligence, which also increase supply chain automation and can help gain a competitive advantage. The supply chain, empowered with the latest technologies, is setting a new paradigm for global competitors. In conjunction with Industry 4.0, it can help in improving the flexibility, agility, and adaptability of the supply chain.

8.1.1 Objectives and Contribution

The objectives of this chapter are:

i. To show the application of Industry 4.0 tools in improving the performance of supply chain management.
ii. To collectivize the issues and challenges arising in the supply chain due to the COVID-19 pandemic.
iii. To present how Industry 4.0 tools can be employed to overcome any disruptions in the supply chain, such as those caused by COVID-19.

This study aims to contribute to the implications of Industry 4.0 tools in the context of enhancing the flexibility of a supply chain and consequently mitigating the complications arising due to unforeseen circumstances, that is, the disruption in supply chain caused by the pandemic.

To further elaborate on these technologies, this literature is structured as follows. Section 8.2 gives an overview of the areas of supply chain and Industry 4.0 along with a review on research works in associated areas. Section 8.3 describes the methodology used to search, review, and analyze research articles. Section 8.4 enlists the issues and challenges arising due to the pandemic in the supply chain management. Section 8.5 elaborates on the different Industry 4.0 tools used to improve performance of a supply chain. Section 8.6 recommends the use of Industry 4.0 tools to overcome the disruptions in a supply chain caused by COVID-19. Section 8.7 concludes the chapter with suggestions on the scope of future research works.

8.2 BACKGROUND

The fourth industrial revolution, also termed Industry 4.0, was coined by the German federal government in 2011 (Gilchrist, 2016). A significant transformation in the digital world was enforced by essential technological tools such as the Internet of Things (IoT), cloud computing, big data, artificial intelligence, and augmented/virtual reality. The exponential growth of advanced technologies and their integration with conventional processes has opened paths to various opportunities for the industrialists to compete. As the customers now desire to verify the authenticity of each part of the product, leveraging Industry 4.0 technologies can enhance traceability and transparency in every stage of the supply chain (Pasi et al., 2020). It also maintains transparency between different supply chain players, including suppliers, logistics providers, manufacturers, retailers, and customers (Calatayud et al., 2016). Industry 4.0 improves the performance of the supply chain by the following (Bär et al., 2018):

i. Integrating all players along the supply chain.
ii. Increasing the degree of customization in a product.
iii. Availability of real-time data across all stakeholders.
iv. Increasing flexibility.
v. Maintaining a transparent and low inventory level.
vi. Minimum level of human interaction in repetitive tasks.
vii. Providing consumer data analytics.

Due to uncertainty in demand, problems of overstocking/understocking, delayed delivery, and so on were prominent in earlier days. In the past few years, risk management in the supply chain is seeking global attention (Ho et al., 2015). As a result, the supply chain has become more susceptible to disruptions with a continuously changing environment, and its long-term effect has been observed (Parast et al., 2019). But the disruption caused by the pandemic has exposed the shortcomings of the conventional supply chain management.

According to a report published by Accenture (Wilson, 2020), supply chain disruptions were observed by 94% of Fortune 1000 companies due to COVID-19; 75% of companies reported to have faced negatively or strongly negative impacts; 55% of companies are planning or have already reduced their expected growth rate. The "Global Economic Prospects" published by The World Bank in June 2020 (Global Economic Prospects, 2020) envisioned the following impact of COVID-19 on the world economy:

i. 5.2% contraction on global GDP in 2020 (the largest dip in decades).
ii. Advanced economies were expected to contract by 7%.
iii. Emerging market and developing economies were forecasted to shrink by 2.5%.

It has been estimated that global businesses are most likely to take more than five years to recover from the current recession caused by the pandemic (Dua et al., 2020). Among them, too, small businesses are likely to shut down or take longer than that to recover. The main cause of businesses failing to recover when faced with such disruptions is a poor recovery strategy (Cerullo & Cerullo, 2004).

A flexible supply chain has much better potential than a rigid supply chain has. It can adjust to complex market patterns, respond effectively in uncertain times, and optimize resource utilization, thereby making the supply chain cost-efficient. However, the supply chain has remained fragile primarily due to manual management, which restricts the operations while dealing with complex processes (Brady, 2020). These complexities are visible at different levels, such as network complexity, process complexity, product complexity, demand complexity, and organizational complexity (Christopher & Holweg, 2011). However, many researchers argue that Industry 4.0 will play a vital role in mitigating problems and improving the flexibility in the supply chain (Shekarian et al., 2020). According to a survey conducted by McKinsey & Company (Agrawal et al., 2021) in more than 400 companies globally, it was reported that 94% of the respondents claimed that Industry 4.0 tools helped them to keep their operations running during the pandemic, and 56% found these tools critical in responding to the crisis.

8.3 METHODOLOGY

The methodology employed was a twofold extensive literature review to identify the implications of Industry 4.0 tools on the supply chain and the challenges posed by the pandemic on the supply chain. The final step was to relate those implications to the challenges, in order to recommend a solution for improving the resilience of the supply chain. The methodology used is shown in Figure 8.1.

The proposed extensive literature review is based on a qualitative analysis of previous research works and is shown in Table 8.1. The aim is to identify Industry 4.0 technologies that enhance supply chain resilience and provide recommendations to face the challenges posed by the COVID-19 pandemic. An extensive literature review was carried out in this order:

Defining the objective: With the aim of understanding the significance of flexible supply chain management in the post-COVID era and analyzing the crucial role that the advanced tools of Industry 4.0 will play, the research objective was framed as an application of Industry 4.0 tools that make a supply chain more resilient. Furthermore, the wide-scale disruptions predominantly in the global supply chain caused by the COVID-19 pandemic makes the use of Industry 4.0 technologies the need of the hour.

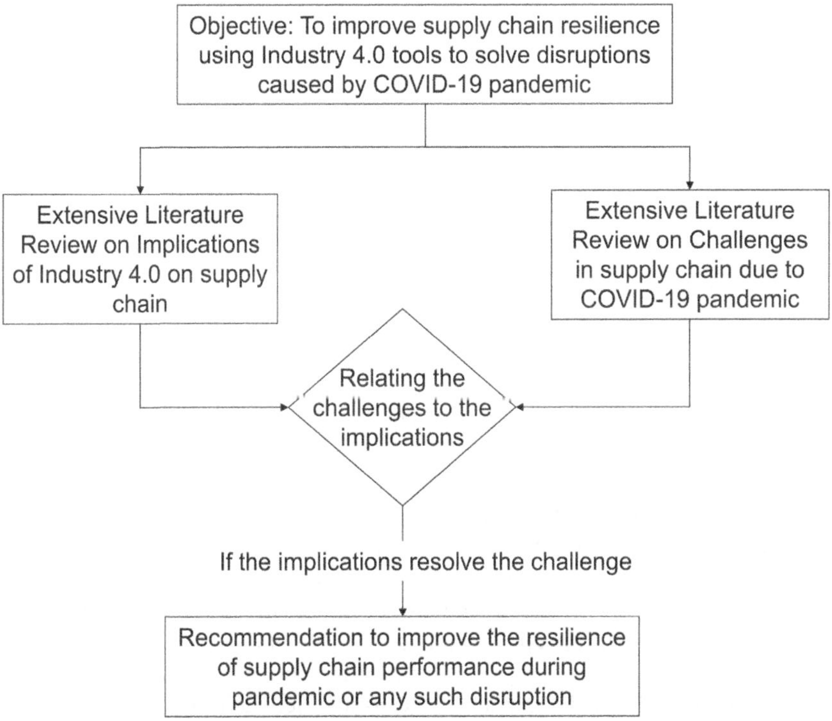

FIGURE 8.1 Methodology used.

TABLE 8.1
Industry 4.0 Tools and Corresponding Relevant Literatures

Big Data	(Blanco et al., 2011); (Rozados & Tjahjono, 2014); (Benabdellah et al., 2016); (Mishra et al., 2016); (Witkowski, 2017); (Roßmann et al., 2018); (Boone et al., 2019); (Q. Li & Liu, 2019); (Vieira et al., 2020a); (Vieira et al., 2020b)
Cloud Computing	(Jun & Wei, 2011); (Helo et al., 2014); (Ko et al., 2016); (Bruque-Cámara et al., 2016); (Han et al., 2017); (Vazquez-Martinez et al., 2018); (Giannakis et al., 2019); (Kong et al., 2020); (S. I. Ali et al., 2021)
Blockchain	(Kshetri, 2018); (Azzi et al., 2019); (Helo & Hao, 2019); (Kawaguchi, 2019); (Batwa & Norrman, 2020); (Dietrich et al., 2020); (Kopyto et al., 2020); (Sund et al., 2020); (Wamba et al., 2020)
Internet of Things	(Tseng et al., 2011); (Wang et al., 2015); (Afsharian et al., 2016); (Verdouw et al., 2016); (Abdel-Basset et al., 2018); (Vallandingham et al., 2018); (Zhou & Piramuthu, 2018); (A. Ali & Haseeb, 2019); (Jiang et al., 2020); (Muñuzuri et al., 2020)
Artificial Intelligence	(Amirkolaii et al., 2017); (Kantasa-ard et al., 2019); (Oleśków-Szłapka et al., 2019)
Augmented Reality	(Cirulis & Ginters, 2013); (Merlino & Sproģe, 2017); (Stoltz et al., 2017); (Mourtzis et al., 2019)

Searching for literature: ScienceDirect (www.sciencedirect.com) and Google Scholar (www.scholar.google.com) were selected as electronic database platforms to get access to a wide range of original and peer-reviewed articles. The keywords were carefully selected, in various combinations of words, such as "Industry 4.0", "supply chain", "flexibility", "COVID-19", "big data", "cloud", "blockchain", "augmented reality", "Internet of Things", "RFID", "wireless network", and "artificial intelligence".

After finding research articles that were close to the scope of this chapter, some technologies were found to be directly related to the challenges identified, while others are yet to be examined for their contribution in recovery of the supply chain from the effects of pandemic. As a result, 45 articles were reviewed under six technologies: Big Data, Cloud Computing, Internet of Things, Blockchain, Artificial Intelligence, and Augmented Reality. Among the literatures reviewed, the majority were on the implications of IoT (22%), Big Data (22%), Cloud Computing, and Blockchain technology (both 20%) as shown in Figure 8.2.

Analyzing the challenges: After reviewing all the challenges due to pandemic and impact of Industry 4.0 tools on supply chain, analysis was made to recommend which technology must be employed to eradicate those challenges. Instead of transforming the entire supply chain at once, it is more feasible to make these changes in phases (Barreto et al., 2017). Moreover, industries are already suffering financially from the impact of the pandemic, such that it will take more than five years to recover from it (Dua et al., 2020). Industry 4.0 technologies are expected to improve the performance of the current supply chain, but this crisis has put new pressure on industries to make changes towards a digital supply chain.

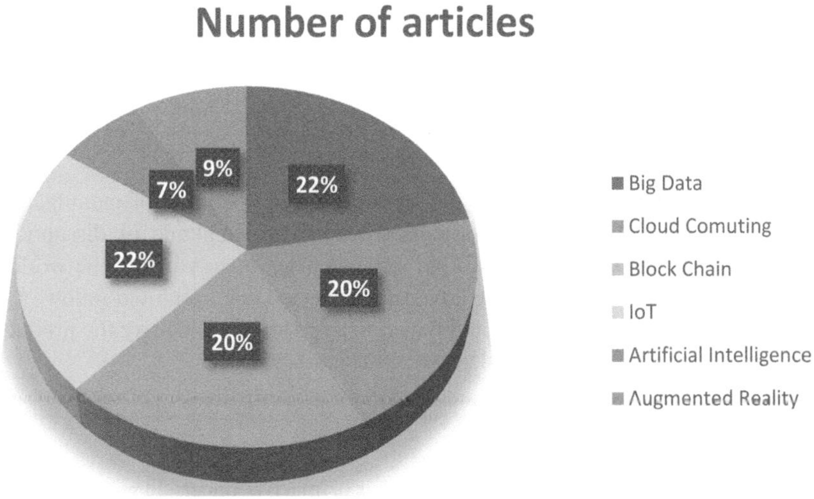

FIGURE 8.2 Distribution of research articles on application of various Industry 4.0 tools in the supply chain.

8.4 ISSUES AND CHALLENGES IN SUPPLY CHAIN DUE TO COVID-19 PANDEMIC

The coronavirus pandemic posed unprecedented challenges to the global supply chain, which compelled many industries to rethink the traditional supply chain models. The majority of impact was caused by the varying prevention policies of various countries and provinces. This section discusses some major impacts of the COVID-19 pandemic on the global supply chain.

8.4.1 Demand

Policies to curb the spread of coronavirus have impacted demand, supply, and logistics in different proportions, with the largest influence on demand at 60%, followed by supply and logistics at 22% and 18% respectively in supply chain aspect (Grida et al., 2020). A sudden change in consumers' preferences tend to have unexpected fluctuations in demand, affecting its critical facets such as forecasting and decision-making process. Likewise, the onset of pandemic had a devastating effect on the global supply chain, which severely disrupted the network (Paul & Chowdhury, 2021). There has been a drastic rise and fall in the demand of essential and non-essential items respectively (Sharma et al., 2020). The essential items such as food and medical equipment (sanitizer, masks, medicines, etc.) witnessed a rise in demand due to panic buying and fear of the unknown, which ultimately led to a shortage of products (Govindan et al., 2020). Managing such high variability in demand becomes difficult especially for small-scale industries, as they do not have the capacity to scale up and the versatility required to absorb demand and supply shocks. Further, many establishments were forced to shut down resulting in profit decline that compelled them to lay off workers in order to cut down costs. Due to a rise in unemployment, consumers avoided unnecessary expenditure and therefore reduced demand greatly for non-essential items (Zhu et al., 2020).

8.4.2 Supply

The lockdown measures restricted the movement and operation of manufacturing plants that remained idle for a long period leading to simultaneous disruption of demand and supply. Companies that are dependent on other parts of the world for procurement of raw materials were adversely impacted due to limited inflow of raw materials. Subsequently, this affected the company's ability to deliver the product to end consumers on time (Tang et al., 2021; Zhu et al., 2020).

8.4.3 Production

In pandemic situations, production disruption is primarily concerned with the unavailability of the workforce and limited production capacity. The number of workforces in industries declined significantly as personnel mobility decreased to abide by the social distancing norms. However, some firms, not relying on labor completely, continued to function and supplied products through online platforms (Mollenkopf et al., 2020). These approaches, however, faced challenges with bottlenecks in sourcing, processing, packaging, and delivery of products. Many companies had insufficient capacity or inventory allocated to e-commerce channels and faced acute challenges corresponding to last-mile delivery (Nordhagen et al., 2021). It implies that each industry should embrace technology-enabled models that can function with minimum workforce and are flexible enough to respond to any unforeseen circumstances in the supply chain. Further, Zhu et al. (2020) throw light on inadequate production capacity of companies with reference to the vaccine supply chain. Several manufacturing companies are competing to develop vaccines. However, the total volume of vaccine required to cover 100% of the world population is considerably large. To meet the demand, companies need to identify the bottlenecks at different phases of the supply chain. Owing to a limited number of manufacturing companies, primarily in the pharmaceutical sector which is the need of the hour, scaling up the production is a key challenge (Alam et al., 2021; Tirivangani et al., 2021).

8.4.4 Logistics

Transportation plays an integral role in keeping the supply chain operational. With lockdowns and quarantine laws imposed globally, businesses struggled to transport raw material or finished goods as aviation and maritime activities were suspended between countries. Even though delivery of indispensable items was permitted, the extended safety measures took a significant amount of time, causing delay and increase in lead time (Xu et al., 2020). While many companies are resorting to an online mode of distribution, transparency to enhance end-to-end supply chain visibility must be prioritized. In response, the global supply chain needs to adopt reliable ways to ensure safe and seamless delivery of essential goods.

8.4.5 Digital Supply Chain

After the impacts of COVID-19, digitalization of the supply chain has become a requisite for proper functioning of businesses. Drawing from an analysis of tweets, Sharma et al. (2020) reveal that not all firms are ready to integrate Industry 4.0 technologies due to lack of basic infrastructure and awareness about Industry 4.0 readiness models. While many companies are adopting advanced technologies, there has also been an escalating concern about cybersecurity and threats across the entire supply chain (Hopkins, 2021). The supply chain is continually evolving with enhanced global connectivity, but at the same time it is vulnerable to threats more than ever. As more organizations are relying on online business, they also need to invest in secured technologies such as blockchain in an effort to ensure improved data security (Weil & Murugesan, 2020).

8.5 INDUSTRY 4.0 TOOLS AND THEIR IMPACT ON FLEXIBILITY OF SUPPLY CHAIN

The potential technologies of Industry 4.0 analyzed in this literature are big data, cloud computing, blockchain, Internet of Things, and augmented reality. Their implications in improving supply chain performance are discussed as follows:

8.5.1 Big Data

Conventionally, the supply chain was assumed to be a linear process that possessed several limitations in dealing with the complex cases of nature and thus failed to respond in time. However, in recent years a large volume of data is generated from the sources associated with a supply chain operation. Its application in business analytics has been continuously increasing with the emergence of information technologies (Mishra et al., 2016; Roßmann et al., 2018). Therefore, effective processing of data is important to make it yielding and reliable (Q. Li & Liu, 2019). While the data processing capability has tremendously improved, it is transforming the supply chain in various dimensions and most significantly in simulation and forecasting (Witkowski, 2017). The data can be collected from various points, such as point-of-source, in-store path data, and user-generated data, in a near real-time system. Subsequently, these data are used to forecast the demand accurately and understand customer preference (Boone et al., 2019). This in turn leads to improvement in supply chain management by enabling autonomous operational and strategic decision-making (Rozados & Tjahjono, 2014). Therefore, it can be established that big data can reduce uncertainty in the supply chain, which simultaneously reduces information processing requirements (Roßmann et al., 2018).

Furthermore, Vieira et al. (2020a) put forth that since traditional simulation approaches are less efficient, supply chain simulation models can significantly benefit from the big data concept when integrated with other Industry 4.0 technologies such as artificial intelligence. The supply chain simulation model captures the dynamic characteristics of the system in real time and can forecast any risk in advance. As a result, various authors have proposed different efficient and flexible methodologies

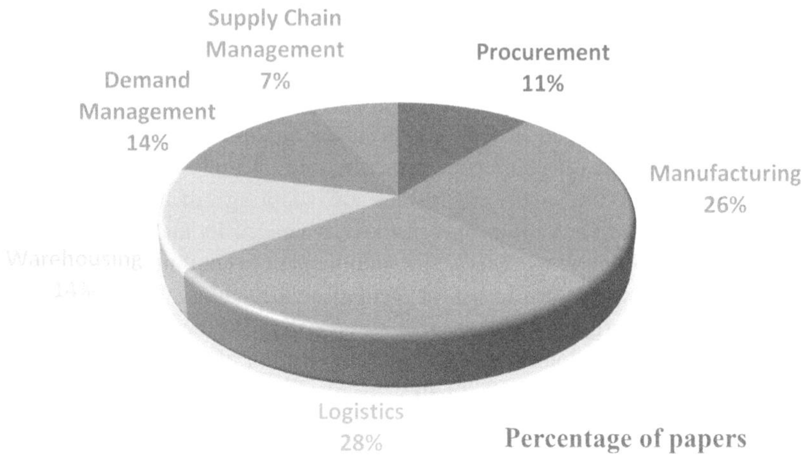

FIGURE 8.3 Percentage of research publications in various supply chain functions in big data analysis.

in tackling different situations within the supply chain (Blanco et al., 2011; Q. Li & Liu, 2019; Vieira et al., 2020b). From the proposed framework by Q. Li and Liu (2019), based on the power split concept for hybrid vehicles, it can be inferred that performing virtual simulation using available data can enhance the overall performance of the supply chain physically. Witkowski (2017) elaborated on the application of big data with an example of DHL, wherein big data technology is integrated into the logistics section of the company. The implementation outcome was found to enhance the operational flexibility and, hence, lead to fulfilling customer obligations. Accessing data from varied sources for optimized decision-making can also improve the competency of logistics. Nguyen et al. (2018) enlisted the percentage of research works in application of big data analysis in various functions of supply chain represented in Figure 8.3.

To increase the adoption of big data in industries, it is also imperative to enlist the challenges faced in this process. Vieira et al. (2020b) registered that those problems mainly were encountered concerning data unavailability. Respective approaches to overcome this barrier were also suggested in an effort to materialize Industry 4.0. It is evident from multiple instances that supply chain complexity can be significantly reduced by leveraging data collection. Thus, companies will respond faster to changing customer needs (Benabdellah et al., 2016).

8.5.2 Cloud Computing

Cloud computing provides an opportunity to connect the entire supply chain operations within a single loop that brings simplicity and ease of accessibility of data throughout the supply chain among all partners. Furthermore, the attributes determining the efficiency of the supply chain are also believed to enhance flexibility (Jun & Wei, 2011). Through different forms of cloud computing services such as

Software as a Service (SaaS), Infrastructure as a Service (IaaS), and Platform as a Service (PaaS), the process of procurement, production, and orders can be coordinated in real time with end-to-end visibility. Cloud computing also facilitates definitive data access for inventory management and proactive risk detection, consequently reducing response time (Giannakis et al., 2019). The ubiquitous cloud computing services that provide mass storage capacity enable the enterprise to utilize resources efficiently, minimize distortion, and improve the speed and accuracy of information sharing via web-based platforms. Bruque-Cámara et al. (2016) and Han et al. (2017) also suggested that integration of cloud computing with the supply chain, which positively affects supply chain performance. A method derived from lean supply chain principles on the cloud platform, namely cloud chain, was formulated to manage the product life cycle. It has increased flexibility since the required changes at any stage of the product life cycle can immediately be made by altering the catalogues along the value chain. The prototype of cloud chain also added to the sovereignty in control and planning of supply chain with secured network settings (Vazquez-Martinez et al., 2018). Cloud-based services can also be used with other technologies to link the product and supply chain design simultaneously. The real-time data obtained can be utilized for personalized decision-making using genetic algorithms (S. I. Ali et al., 2021). Due to high implementation costs, mainly IT-based large enterprises employ solutions. However, Ko et al. (2016) has developed a cost-efficient cloud-based system for small-to-medium business that makes them globally competitive by increasing transparency and flexibility in the supply chain. Another prevalent technology is the enterprise resource planning (ERP) system which has evolved; however, there are limitations, such as continuously changing consumer needs and lack of adaptability, which can be overcome by cloud services (Helo et al., 2014). The need for flexibility in the supply chain led to the formation of CPS-enabled multi-layer cloud solutions to reduce logistical complexities. For example, there are cloud-based decentralized wearable assets that are capable of managing logistics and synchronization tools for notifying pickup and delivery status, which optimize resource utilization (Kong et al., 2020).

8.5.3 Blockchain Technology

A relatively new concept for manufacturing industries, blockchain was first introduced in 2009, by Satoshi Nakamoto, who also created one of the cryptocurrencies, Bitcoin (Kawaguchi, 2019). As opposed to maintaining individual ledgers, blockchain technology stores data in blockchains accessible to every stakeholder, increasing transparency and trust between parties (shown in Figure 8.4). The blockchain is highly secured with a cryptographic signature which prevents data manipulation in retrospect and minimizes other cybersecurity threats (Azzi et al., 2019). Blockchain architecture is prepared to verify transactions to decentralize the logistics system and increase flexibility and automation. At the same time, different actors of the supply chain can use functions provided by this technology to record history and approve orders (Batwa & Norrman, 2020). It also brings flexibility in handling complex situations due to its concept of distributed transaction ledgers (Helo & Hao, 2019). In the same context, Kawaguchi (2019) proposed a blockchain model incorporating

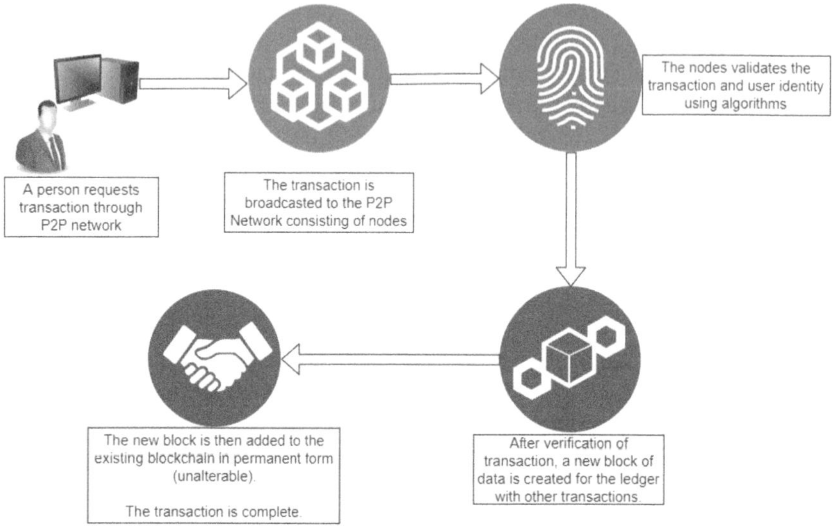

FIGURE 8.4 Steps involved in blockchain transaction.

distributed storage. There is an interplanetary file system that transfers data safely and increases storage capacity for the large volume of data available.

Sund et al. (2020) discussed a prototype with traceability of data using blockchain technology, expanding the supply chain event processing efficiency. About the supply chain of a manufacturing industry, Dietrich et al. (2020) proposed a framework wherein each component of an individual product is identified by a unique identification number generated from blockchain technology, which would enhance traceability of defective parts. Kshetri (2018) put forth that the food industry, which involves multiple parties in its supply chain network, is most likely to be impacted by blockchain. The virtual identity ensures traceability, such as tracking a defective product's origin and calling the entire batch back in the minimum period. This increases auditability, which in turn elucidates supply chain transparency and flexibility. The findings of the empirical study by Wamba et al. (2020) shed light on the advantages of blockchain in supply chain management, and it can be inferred that blockchain plays a vital role in knowledge sharing, which is significant for performing the work flexibility. Smart contracts, one of the most efficient components of blockchain technology, are anticipated to substitute operational tasks by the year 2035 (Kopyto et al., 2020). Nevertheless, automation cannot be achieved entirely for non-standard supply chain processes due to its high level of complexity. The research in blockchain application in the supply chain is still in its nascent stage (Wamba et al., 2020).

8.5.4 Internet of Things (IoT)

Due to the earlier unavailability of data in real time, inaccuracies in the supply chain were unavoidable, which resulted in cost and workforce wastage. The IoT application has overcome this, and its further integration will boost the reconfiguration of existing supply chain models. Verdouw et al. (2016) addressed supply chain complexity attributes to four factors: network, object, process, and control. These complexities can be simplified by virtualizing the entire supply chain process. IoT enables self-adaptability in supply chains, with benefits such as intelligent placement of orders, reduced lead time, and minimized operational cost. It is also used in designing control systems to optimize and monitor intermodal transportation that provides improved command over inventory control and flexibility by virtually modifying the transport plan (Muñuzuri et al., 2020). These developments have also been proved to reduce overhead costs. The most widely used IoT technologies by industries are radio-frequency identification (RFID) and wireless sensor network (WSN), which are further discussed in this section.

8.5.4.1 Radio-Frequency Identification (RFID)

RFID tags are progressively being used to generate data from various sources for maintaining inventory at all levels and automatic scheduling (Afsharian et al., 2016; A. Ali & Haseeb, 2019). Zhou and Piramuthu (2018) proposed a framework based on RFID to handle technological aspects and minimize communication barriers by efficiently utilizing the data for seamless business operation. About a grocery retail supply chain, Vallandingham et al. (2018) state that by integrating RFID and wireless sensors, planning and control can be executed with a high level of accuracy and precision. For perishable products, the real-time temperature can be obtained with end-to-end visibility, which subsequently maintains the quality of the product. Another important aspect is that it reduces waste and makes planning flexible and dynamic. However, using RFID alone is not sufficient and requires other enabling technology to verify the result. Abdel-Basset et al. (2018) have modeled a framework to enhance inventory management, real-time supply chain management, and transparency in logistics. On scanning each product installed with RFID tags, a database is created containing information such as date of production and expiry and warranty period.

8.5.4.2 Wireless Sensor Network (WSN)

Another critical enabler in the IoT paradigm is the wireless sensor network that can monitor the logistics system by combining big data and sensors (Jiang et al., 2020). The advancement in microelectronics has led to a significant reduction in the size of sensor nodes that can be installed inside perishable product packaging for tracking parameters that affect the quality of product (Wang et al., 2015). As researchers continue to explore the possibilities of using WSN to consume minimum power, a new framework was developed to monitor mechanical vibrations or shocks causing damage due to unforeseen incidents; for example, falling of product due to vibration in a vehicle or an accident. Through the extended reach of IoT, the data recorded by sensors can be obtained on smartphones and PCs of the end user, thereby alerting them in unusual situations (Tseng et al., 2011).

Other technologies contributing to IoT are the global positioning system (GPS) and geographic information system (GIS). The comprehensive information generated by the spatially distributed network that connects devices with embedded autonomous

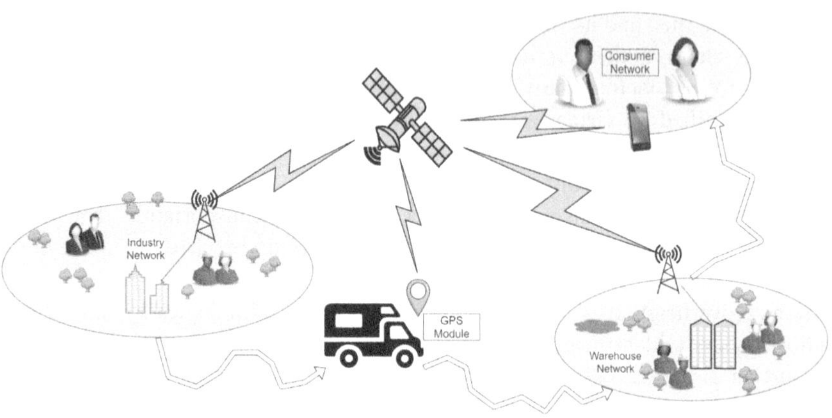

FIGURE 8.5 A supply chain network with GPS-enabled real-time tracking.

sensors opens up new dimensions to explore (as shown in Figure 8.5), making the supply chain reconfigurable and thus more resilient.

8.5.5 Artificial Intelligence

The growing application of artificial intelligence (AI) drives the supply chain towards more flexible execution and efficient handling of interconnected networks. AI can mitigate the risk posed by unexpected events disrupting the supply chain by using a large volume of data generated and consequently improving the decision-making process (Oleśków-Szłapka et al., 2019). To keep up with the erratic nature of demand patterns and changing consumer expectations, AI ensures accurate forecasting and reduces inventory management costs. The existing forecasting method that depends on the linear or fixed trajectory is insufficient due to high uncertainty in demand and volatility in the market. Methods based on AI produced more accurate results of forecasting, which was validated by an aircraft company (Amirkolaii et al., 2017). Advanced technologies combined with AI enable flexible operation of logistics and distribution centers by autonomous handling of complex situations. For example, Kantasa-ard et al. (2019) analyzed variations in demand, based on a case study of white sugar consumption, using a neural network model, wherein the proposed model positively affected the production capacity and inventory management. However, more research work needs to be focused on AI in the supply chain context to verify its contribution towards flexibility.

8.5.6 Augmented Reality

Beyond the experience of using virtual 3D projection in video games, augmented reality (AR) developments are now being recognized by companies to ameliorate supply chain operations. It is being implemented in almost every field, but logistics is relatively a newer dimension to explore these possibilities (Cirulis & Ginters,

2013). The traditional method of paper-based verification required when picking and delivering orders is susceptible to human error, leading to a significant loss for the organization. The process is proved to be non-productive and is now being gradually replaced by devices that can scan the product through a camera-operated system. After DHL introduced AR in its supply chain, the outcome transformed the ways of working, specifically for the labor-intensive works, in various aspects such as optimizing the picking/lifting operation. It provides the most efficient path to the nearest package within the warehouse in an AR environment. With the help of wearables available such as heads-up displays, like Google Lens, the instructions to be followed by the operators are displayed in these devices to be able to see quickly and follow. AR can increase overall efficiency by addressing the challenges that include poor resource utilization, inadequate planning, and scheduling (Merlino & Sproġe, 2017). As AR has enhanced the visualization, bridging the gap between the real and virtual worlds, it has reduced warehouse management complexities and optimized logistics costs. The implementation of smart glasses supported by software allows flexible navigating through the warehouse following optimized routes (Mourtzis et al., 2019). It improves the accuracy and productivity of workers with a minimum training period. DHL observed a 25% increase in efficiency as order picking speed had significantly improved (Schrauf & Berttram, 2016). AR provides a remote connection between various players of the supply chain in a virtual environment. For example, an operator need not be physically present to verify the product; hence, irrespective of time and location, any information can be transferred (Stoltz et al., 2017). Consequently, it leads to greater flexibility and agility in the supply chain. However, technological limitations, such as the system's incompatibility with the existing infrastructure and use of non-standard software, make the installation of AR technology expensive (Ginters & Martin-Gutierrez, 2013; Stoltz et al., 2017).

8.6 RECOMMENDATIONS

This section elaborates on the recommendations about the use of Industry 4.0 tools that can assist in overcoming the challenges created by the COVID-19 pandemic. The recommendations are divided into five major areas in which the challenges were identified: demand, supply, production, logistics, and digital supply chain.

8.6.1 Demand

On the demand side of the supply chain, some major problems encountered were in the area of forecasting and decision-making process. The use of conventional tools to forecast highly fluctuating demands are often unreliable and result in understock or overstock inventory. In order to prevent financial losses due to misleading analysis, big data can play a major role in the immediate response to the COVID-19 pandemic. Firms can utilize the vast amount of data for real-time decision-making to align with changing consumer demands. Further, a dedicated system equipped with advanced software can be designed by integrating IoTs with tools such as artificial intelligence and big data to automatically place orders when the need arises by understanding the pattern of change in demand. In a post-COVID era, these tools will support professionals in preventing future supply chain disruptions.

8.6.2 Supply

The mobility restrictions imposed by the government of different countries in light of the COVID-19 pandemic was more disadvantageous for firms dependent on a limited number of suppliers. However, the application of cloud computing enables them to connect globally, and hence the probability of inflow of raw material increases. Cloud-based services also assist in improving communication through information flow and real-time connectivity between different supply chain actors. This is all the more helpful as more customers are now disposed to gain information at every stage of the supply chain. These technologies have positive implications on supply chain resilience, and in addition, functional and economic efficiency of the organization is enhanced.

8.6.3 Production

Productivity of small as well as large firms has been greatly disrupted by the sudden impact of COVID-19. This has made industrialists realize the need for automation and digitization of manufacturing processes to sustain them when facing obstructions in the future. To overcome the shortage of workforce, artificial intelligence along with IoT should be employed. Advanced machines are able to perform manual and repetitive work with minimum technical assistance and produce better quality products. Leveraging these technologies, companies can keep their manufacturing unit functional during pandemic-like situations and provide a safe environment for their workers, in addition to improving production efficiency. Artificial intelligence can positively affect the production capacity, wherein the company is prepared to meet the customer's demand in any circumstance, such as in case of the COVID-19 pandemic. Moreover, automation also adds flexibility to the supply chain as evident from the fact that some automobile companies were able to transform their production line to manufacture ventilators in times of crisis.

8.6.4 Logistics

Augmented reality can be highly useful in the times of COVID-19 outbreak, as people are required to maintain proper physical distance. The companies can use them for contactless delivery of products to their customers, which is also less time-consuming. Another important tool is IoT that facilitates tracking of products at every stage and thus creating end-to-end visibility throughout the supply chain without being present physically. For example, vaccine temperature can be easily monitored and controlled with the help of wireless sensors and GPS systems, considerably reducing vaccine wastage. Use of these sophisticated tools can overcome the challenges of last-mile delivery and at the same time minimize the transportation cost. Improving logistics efficiency is one of the crucial strategies to incorporate resilience in the supply chain network.

8.6.5 Digital Supply Chain

Amidst technological advancement, the supply chain system is exposed to a high level of vulnerability whereby company's sensitive data, which should be protected to ensure privacy, are at risk. To overcome these challenges, blockchain technology provides a secure platform for storing data and transactions. Due to blockchain's decentralized network, a high degree of transparency and traceability can be achieved globally, as a result making the supply chain resilient. These characteristics of the supply chain are essential to conduct safe business especially in the post-pandemic period when a large number of companies have shifted to online bases. According to a survey, 90% of top-level executives have emphasized the importance of blockchain in the post-COVID era (Gupta et al., 2020). It provides an alternate to the traditional work technique and a safe environment that enables working remotely to prevent further COVID-19 outbreaks.

8.7 DISCUSSION

Industry 4.0 technologies have different applications within the supply chain that contribute to transparency and better collaboration of supply chain actors. With several potential benefits in store, such as better inventory management, decreased staffing leading to reduced error, end-to-end visibility for better communication and cost-efficient system, these technologies contribute to a flexible supply chain that is anticipated to be competent enough to address forthcoming disruptions.

Big data and AI allow process optimization and accurate prediction to meet volatile demand, while cloud computing, IoT, and AR increase the interconnectivity of supply chain operations. RFID and WSN improve automation by collecting information about the product, and subsequently, they make logistics more flexible and intelligent. Blockchain and IoT, when used together, provide secured transactions and better traceability along the supply chain process. In addition, they also promote security within the system; for example, the blockchain provides an immutable ledger that is cryptographically protected.

Moreover, it was observed that the advanced technologies of Industry 4.0, employed in warehouses, delivery centers, and logistics, enabled real-time visibility and accessibility of information and reduced lead time and operational cost, which simultaneously improved customer satisfaction (Gilchrist, 2016; Oh & Jeong, 2018).

Some of the reviewed papers also identified the technological limitations since these technologies are still in their developing stages (Horváth & Szabó, 2019; Mishra et al., 2016; Romportl, 2015). However, the continuing pace of advancement in research focusing on the reconfiguration of the supply chain will lead to faster adoption of these tools.

8.8 CONCLUSION

The coronavirus pandemic led to a devastating impact on the global supply chain and posed unique challenges for industrialists. The technologies of Industry 4.0 such as big data, artificial intelligence, cloud computing, blockchain, IoT, and augmented reality provide innovative solutions and produce improved results that support the requirements of the supply chain in the post-COVID era. It is conceivable that by leveraging Industry 4.0 technologies, companies will be more adaptive to the rapidly changing environment in future.

The following scopes for future research areas have been found out to explore possible avenues related to supply chain flexibility, with these advanced technologies growing exponentially:

1. Quantifiable impact of technologies in terms of supply chain performance should be studied to enhance planning and control of the retail supply chain (Vallandingham et al., 2018).
2. More blockchain-based models should be tested in the field of the supply chain, since its decentralized nature and capability to analyze and forecast can help to a great extent in enhancing flexibility.
3. Studies should also focus on analyzing the economic impact of advanced technologies considering their implementation cost for different applications in the supply chain.
4. The drawbacks and barriers in adopting these technologies should be identified to develop a flexible supply chain.

As these technologies are readily available, more companies can benefit from their comprehensive use through which the supply chain can achieve high levels of transparency, flexibility, and resilience, subsequently improving the overall performance of the company. The long-term effects of pandemic are still unknown; however, prompt action by an organization can significantly mitigate future disruptions.

REFERENCES

Abdel-Basset, M., Manogaran, G., & Mohamed, M. (2018). Internet of Things (IoT) and its impact on supply chain: A framework for building smart, secure and efficient systems. *Future Generation Computer Systems*, *86*, 614–628. https://doi.org/10.1016/j.future.2018.04.051

Afsharian, S. P., Alizadeh, A., & Chehrehpak, M. (2016). Effects of applying radio frequency identification in supply chain management: An empirical study of manufacturing enterprises. *International Journal of Business Information Systems*, *23*(1), 97. https://doi.org/10.1504/IJBIS.2016.078026

Agrawal, M., Dutta, S., Kelly, R., & Millán, I. (2021). *How the Pandemic Transformed Digital Manufacturing – and Vice Versa | McKinsey*. McKinsey & Company. https://www.mckinsey.com/business-functions/operations/our-insights/covid-19-an-inflection-point-for-industry-40#

Alam, S. T., Ahmed, S., Ali, S. M., Sarker, S., Kabir, G., & Ul-Islam, A. (2021). Challenges to COVID-19 vaccine supply chain: Implications for sustainable development goals. *International Journal of Production Economics*, *239*, 108193. https://doi.org/10.1016/j.ijpe.2021.108193

Ali, A., & Haseeb, M. (2019). Radio frequency identification (RFID) technology as a strategic tool towards higher performance of supply chain operations in textile and apparel industry of Malaysia. *Uncertain Supply Chain Management*, 215–226. https://doi.org/10.5267/j.uscm.2018.10.004

Ali, S. I., Ali, A., AlKilabi, M., & Christie, M. (2021). Optimal supply chain design with product family: A cloud-based framework with real-time data consideration. *Computers & Operations Research, 126*, 105112. https://doi.org/10.1016/j.cor.2020.105112

Amirkolaii, K. N., Baboli, A., Shahzad, M. K., & Tonadre, R. (2017). Demand forecasting for irregular demands in business aircraft spare parts supply chains by using Artificial Intelligence (AI). *IFAC-PapersOnLine, 50*(1), 15221–15226. https://doi.org/10.1016/j.ifacol.2017.08.2371

Azzi, R., Chamoun, R. K., & Sokhn, M. (2019). The power of a blockchain-based supply chain. *Computers & Industrial Engineering, 135*, 582–592. https://doi.org/10.1016/j.cie.2019.06.042

Bär, K., Nadja, Z., Herbert, L., & Waqas, H. (2018). Considering Industry 4. 0 aspects in the supply chain for an SME. *Production Engineering, 0*(0), 0. https://doi.org/10.1007/s11740-018-0851-y

Barreto, L., Amaral, A., & Pereira, T. (2017). Industry 4.0 implications in logistics: An overview. *Procedia Manufacturing, 13*(December), 1245–1252. https://doi.org/10.1016/j.promfg.2017.09.045

Batwa, A., & Norrman, A. (2020). A framework for exploring blockchain technology in supply chain management. *Operations and Supply Chain Management: An International Journal*, 294–306. https://doi.org/10.31387/oscm0420271

Benabdellah, A. C., Benghabrit, A., Bouhaddou, I., & Zemmouri, E. M. (2016). Big data for supply chain management: Opportunities and challenges. *2016 IEEE/ACS 13th International Conference of Computer Systems and Applications (AICCSA)*, 1–6. https://doi.org/10.1109/AICCSA.2016.7945828

Bertrand, J. W. M. (2003). *Supply Chain Design: Flexibility Considerations* (pp. 133–198). https://doi.org/10.1016/S0927-0507(03)11004-3

Blanco, E. E., Xu Yang, Gralla, E., Godding, G., & Rodriguez, E. (2011). Using discrete-event simulation for evaluating non-linear supply chain phenomena. *Proceedings of the 2011 Winter Simulation Conference (WSC)*, 2255–2267. https://doi.org/10.1109/WSC.2011.6147937

Boone, T., Ganeshan, R., Jain, A., & Sanders, N. R. (2019). Forecasting sales in the supply chain: Consumer analytics in the big data era. *International Journal of Forecasting, 35*(1), 170–180. https://doi.org/10.1016/j.ijforecast.2018.09.003

Brady, D. (2020). *COVID-19 and Supply-chain Recovery: Planning for the Future | McKinsey*. McKinsey & Company. https://www.mckinsey.com/business-functions/operations/our-insights/covid-19-and-supply-chain-recovery-planning-for-the-future#

Bruque-Cámara, S., Moyano-Fuentes, J., & Maqueira-Marín, J. M. (2016). Supply chain integration through community cloud: Effects on operational performance. *Journal of Purchasing and Supply Management, 22*(2), 141–153. https://doi.org/10.1016/j.pursup.2016.04.003

Burin, A. R. G., Perez-Arostegui, M. N., & Llorens-Montes, J. (2020). Ambidexterity and IT competence can improve supply chain flexibility? A resource orchestration approach. *Journal of Purchasing and Supply Management, 26*(2), 100610. https://doi.org/10.1016/j.pursup.2020.100610

Calatayud, A., Palacin, R., Mangan, J., Jackson, E., & Ruiz-Rua, A. (2016). Understanding connectivity to international markets: A systematic review. *Transport Reviews, 36*(6), 713–736. https://doi.org/10.1080/01441647.2016.1157836

Cerullo, V., & Cerullo, M. J. (2004). Business continuity planning: A comprehensive approach. *Information Systems Management, 21*(3), 70–78. https://doi.org/10.1201/1078/44432.21.3.20040601/82480.11

Christopher, M., & Holweg, M. (2011). "Supply Chain 2.0": Managing supply chains in the era of turbulence. *International Journal of Physical Distribution & Logistics Management, 41*(1), 63–82. https://doi.org/10.1108/09600031111101439

Cirulis, A., & Ginters, E. (2013). Augmented reality in logistics. *Procedia Computer Science, 26*, 14–20. https://doi.org/10.1016/j.procs.2013.12.003

Dietrich, F., Palm, D., & Louw, L. (2020). Smart contract based framework to increase transparency of manufacturing networks. *Procedia CIRP, 91*, 278–283. https://doi.org/10.1016/j.procir.2020.02.177

Dua, A., Mahajan, D., Oyer, L., & Ramaswamy, S. (2020). *US Small-business Recovery after the COVID-19 Crisis | McKinsey*. McKinsey & Company. https://www.mckinsey.com/industries/public-and-social-sector/our-insights/us-small-business-recovery-after-the-covid-19-crisis

Eckstein, D., Goellner, M., Blome, C., & Henke, M. (2015). The performance impact of supply chain agility and supply chain adaptability: The moderating effect of product complexity. *International Journal of Production Research, 53*(10), 3028–3046. https://doi.org/10.1080/00207543.2014.970707

Giannakis, M., Spanaki, K., & Dubey, R. (2019). A cloud-based supply chain management system: Effects on supply chain responsiveness. *Journal of Enterprise Information Management, 32*(4), 585–607. https://doi.org/10.1108/JEIM-05-2018-0106

Gilchrist, A. (2016). Introducing industry 4.0. In *Industry 4.0* (pp. 195–215). https://doi.org/10.1007/978-1-4842-2047-4

Ginters, E., & Martin-Gutierrez, J. (2013). Low Cost Augmented Reality and RFID Application for Logistics Items Visualization. *Procedia Computer Science, 26*, 3–13. https://doi.org/10.1016/j.procs.2013.12.002

Global Economic Prospects. (2020). *The Global Economic Outlook During the COVID-19 Pandemic: A Changed World*. The World Bank. https://www.worldbank.org/en/news/feature/2020/06/08/the-global-economic-outlook-during-the-covid-19-pandemic-a-changed-world

Govindan, K., Mina, H., & Alavi, B. (2020). A decision support system for demand management in healthcare supply chains considering the epidemic outbreaks: A case study of coronavirus disease 2019 (COVID-19). *Transportation Research Part E: Logistics and Transportation Review, 138*(April), 101967. https://doi.org/10.1016/j.tre.2020.101967

Grida, M., Mohamed, R., & Zaied, A. N. H. (2020). Evaluate the impact of COVID-19 prevention policies on supply chain aspects under uncertainty. *Transportation Research Interdisciplinary Perspectives, 8*, 100240. https://doi.org/10.1016/j.trip.2020.100240

Gupta, S., Snowdon, J., & Mondal, tanmoy. (2020). State of enterprise blockchain marker 2020. In *HFS Research Ltd*. https://www.wipro.com/content/dam/nexus/en/analyst-speak/pdfs/state-of-enterprise-blockchain-market-2020-new.pdf

Han, J. H., Wang, Y., & Naim, M. (2017). Reconceptualization of information technology flexibility for supply chain management: An empirical study. *International Journal of Production Economics, 187*, 196–215. https://doi.org/10.1016/j.ijpe.2017.02.018

Helo, P., & Hao, Y. (2019). Blockchains in operations and supply chains: A model and reference implementation. *Computers & Industrial Engineering, 136*, 242–251. https://doi.org/10.1016/j.cie.2019.07.023

Helo, P., Suorsa, M., Hao, Y., & Anussornnitisarn, P. (2014). Toward a cloud-based manufacturing execution system for distributed manufacturing. *Computers in Industry, 65*(4), 646–656. https://doi.org/10.1016/j.compind.2014.01.015

Ho, W., Zheng, T., Yildiz, H., & Talluri, S. (2015). Supply chain risk management: A literature review. *International Journal of Production Research, 53*(16), 5031–5069. https://doi.org/10.1080/00207543.2015.1030467

Hopkins, J. L. (2021). An investigation into emerging industry 4.0 technologies as drivers of supply chain innovation in Australia. *Computers in Industry, 125*, 103323. https://doi.org/10.1016/j.compind.2020.103323

Horváth, D., & Szabó, R. Z. (2019). Driving forces and barriers of Industry 4.0: Do multinational and small and medium-sized companies have equal opportunities? *Technological Forecasting and Social Change*, *146*(October 2018), 119–132. https://doi.org/10.1016/j.techfore.2019.05.021

Jiang, J., Wang, H., Mu, X., & Guan, S. (2020). Logistics industry monitoring system based on wireless sensor network platform. *Computer Communications*, *155*, 58–65. https://doi.org/10.1016/j.comcom.2020.03.016

Jun, C., & Wei, M. Y. (2011). The research of supply chain information collaboration based on cloud computing. *Procedia Environmental Sciences*, *10*, 875–880. https://doi.org/10.1016/j.proenv.2011.09.140

Kantasa-ard, A., Bekrar, A., el cadi, A. A., & Sallez, Y. (2019). Artificial intelligence for forecasting in supply chain management: A case study of White Sugar consumption rate in Thailand. *IFAC-PapersOnLine*, *52*(13), 725–730. https://doi.org/10.1016/j.ifacol.2019.11.201

Kawaguchi, N. (2019). Application of blockchain to supply chain: Flexible blockchain technology. *Procedia Computer Science*, *164*, 143–148. https://doi.org/10.1016/j.procs.2019.12.166

Ko, H. S., Azambuja, M., & Felix Lee, H. (2016). Cloud-based materials tracking system prototype integrated with radio frequency identification tagging technology. *Automation in Construction*, *63*, 144–154. https://doi.org/10.1016/j.autcon.2015.12.011

Kong, X. T. R., Zhong, R. Y., Zhao, Z., Shao, S., Li, M., Lin, P., Chen, Y., Wu, W., Shen, L., Yu, Y., & Huang, G. Q. (2020). Cyber physical ecommerce logistics system: An implementation case in Hong Kong. *Computers & Industrial Engineering*, *139*, 106170. https://doi.org/10.1016/j.cie.2019.106170

Kopyto, M., Lechler, S., von der Gracht, H. A., & Hartmann, E. (2020). Potentials of blockchain technology in supply chain management: Long-term judgments of an international expert panel. *Technological Forecasting and Social Change*, *161*, 120330. https://doi.org/10.1016/j.techfore.2020.120330

Kshetri, N. (2018). 1 Blockchain's roles in meeting key supply chain management objectives. *International Journal of Information Management*, *39*, 80–89. https://doi.org/10.1016/j.ijinfomgt.2017.12.005

Li, Q., & Liu, A. (2019). Big data driven supply chain management. *Procedia CIRP*, *81*, 1089–1094. https://doi.org/10.1016/j.procir.2019.03.258

Li, X., Goldsby, T. J., & Holsapple, C. W. (2009). Supply chain agility: Scale development. *The International Journal of Logistics Management*, *20*(3), 408–424. https://doi.org/10.1108/09574090911002841

Merlino, M., & Sproģe, I. (2017). The augmented supply chain. *Procedia Engineering*, *178*, 308–318. https://doi.org/10.1016/j.proeng.2017.01.053

Mishra, D., Gunasekaran, A., Papadopoulos, T., & Childe, S. J. (2016). Big Data and supply chain management : A review and bibliometric analysis. *Annals of Operations Research*. https://doi.org/10.1007/s10479-016-2236-y

Mollenkopf, D. A., Ozanne, L. K., & Stolze, H. J. (2020). A transformative supply chain response to COVID-19. *Journal of Service Management*, *32*(2), 190–202. https://doi.org/10.1108/JOSM-05-2020-0143

Mourtzis, D., Samothrakis, V., Zogopoulos, V., & Vlachou, E. (2019). Warehouse design and operation using augmented reality technology: A papermaking industry case study. *Procedia CIRP*, *79*, 574–579. https://doi.org/10.1016/j.procir.2019.02.097

Muñuzuri, J., Onieva, L., Cortés, P., & Guadix, J. (2020). Using IoT data and applications to improve port-based intermodal supply chains. *Computers and Industrial Engineering*, *139*(xxxx), 105668. https://doi.org/10.1016/j.cie.2019.01.042

Nguyen, T., ZHOU, L., Spiegler, V., Ieromonachou, P., & Lin, Y. (2018). Big data analytics in supply chain management: A state-of-the-art literature review. *Computers and Operations Research*, *98*, 254–264. https://doi.org/10.1016/j.cor.2017.07.004

Nordhagen, S., Igbeka, U., Rowlands, H., Shine, R. S., Heneghan, E., & Tench, J. (2021). COVID-19 and small enterprises in the food supply chain: Early impacts and implications for longer-term food system resilience in low- and middle-income countries. *World Development*, *141*, 105405. https://doi.org/10.1016/j.worlddev.2021.105405

Oh, J., & Jeong, B. (2018). Tactical supply planning in smart manufacturing supply chain. *Robotics and Computer-Integrated Manufacturing*, *April 2017*, 0–1. https://doi.org/10.1016/j.rcim.2018.04.003

Oleśków-Szłapka, J., Wojciechowski, H., Domański, R., & Pawłowski, G. (2019). Logistics 4.0 maturity levels assessed based on GDM (grey decision model) and artificial intelligence in logistics 4.0 -trends and future perspective. *Procedia Manufacturing*, *39*, 1734–1742. https://doi.org/10.1016/j.promfg.2020.01.266

Parast, M. M., Sabahi, S., & Kamalahmadi, M. (2019). *The Relationship Between Firm Resilience to Supply Chain Disruptions and Firm Innovation* (pp. 279–298). https://doi.org/10.1007/978-3-030-03813-7_17

Pasi, B. N., Mahajan, S. K., & Rane, S. B. (2020). Smart supply chain management: A perspective of industry 4.0 smart supply chain management: A perspective of industry 4.0. *International Journal of Advanced Science and Technology*, *29*(5), 3016–3030. https://doi.org/10.13140/RG.2.2.29012.01920

Paul, S. K., & Chowdhury, P. (2021). A production recovery plan in manufacturing supply chains for a high-demand item during COVID-19. *International Journal of Physical Distribution and Logistics Management*, *51*(2), 104–125. https://doi.org/10.1108/IJPDLM-04-2020-0127

Ranney, M. L., Griffeth, V., & Jha, A. K. (2020). Critical supply shortages – the need for ventilators and personal protective equipment during the Covid-19 pandemic. *New England Journal of Medicine*, *382*(18), e41. https://doi.org/10.1056/NEJMp2006141

Romportl, J. (2015). *Beyond Artificial Intelligence: The Disappearing Human-Machine Divide* (Vol. 9). https://doi.org/10.1007/978-3-319-09668-1

Roßmann, B., Canzaniello, A., von der Gracht, H., & Hartmann, E. (2018). The future and social impact of big data analytics in supply chain management: Results from a Delphi study. *Technological Forecasting and Social Change*, *130*(September), 135–149. https://doi.org/10.1016/j.techfore.2017.10.005

Rozados, I. V., & Tjahjono, B. (2014). Big data analytics in supply chain management: Trends and related research. *6th International Conference on Operations and Supply Chain Management, Bali, 2014*, *1*(1), 2013–2014. https://doi.org/10.13140/RG.2.1.4935.2563

Schrauf, S., & Berttram, P. (2016). *Indsutry 4.0: How Digitization Makes the Supply Chain More Efficient, Agile, and Customer-focused*. https://www.strategyand.pwc.com/gx/en/insights/2016/digitization-more-efficient.html

Sharma, A., Adhikary, A., & Borah, S. B. (2020). Covid-19's impact on supply chain decisions: Strategic insights from NASDAQ 100 firms using Twitter data. *Journal of Business Research*, *117*, 443–449. https://doi.org/10.1016/j.jbusres.2020.05.035

Shekarian, M., Reza Nooraie, S. V., & Parast, M. M. (2020). An examination of the impact of flexibility and agility on mitigating supply chain disruptions. *International Journal of Production Economics*, *220*, 107438. https://doi.org/10.1016/j.ijpe.2019.07.011

Sherman, E. (2020). *Coronavirus Impact: 94% of the Fortune 1000 are Seeing Supply Chain Disruptions | Fortune*. Fortune. https://fortune.com/2020/02/21/fortune-1000-coronavirus-china-supply-chain-impact/

Stoltz, M.-H., Giannikas, V., McFarlane, D., Strachan, J., Um, J., & Srinivasan, R. (2017). Augmented reality in warehouse operations: Opportunities and barriers. *IFAC-PapersOnLine*, *50*(1), 12979–12984. https://doi.org/10.1016/j.ifacol.2017.08.1807

Sund, T., Lööf, C., Nadjm-Tehrani, S., & Asplund, M. (2020). Blockchain-based event processing in supply chains – A case study at IKEA. *Robotics and Computer-Integrated Manufacturing*, *65*, 101971. https://doi.org/10.1016/j.rcim.2020.101971

Tang, C.-H. H., Chin, C.-Y. Y., & Lee, Y.-H. H. (2021). Coronavirus disease outbreak and supply chain disruption: Evidence from Taiwanese firms in China. *Research in International Business and Finance*, *56*(October 2020), 101355. https://doi.org/10.1016/j.ribaf.2020.101355

Tirivangani, T., Alpo, B., Kibuule, D., Gaeseb, J., & Adenuga, B. A. (2021). Impact of COVID-19 pandemic on pharmaceutical systems and supply chain – a phenomenological study. *Exploratory Research in Clinical and Social Pharmacy*, *2*, 100037. https://doi.org/10.1016/j.rcsop.2021.100037

Tseng, M.-L., Wu, K.-J., & Nguyen, T. T. (2011). Information technology in supply chain management: A case study. *Procedia – Social and Behavioral Sciences*, *25*, 257–272. https://doi.org/10.1016/j.sbspro.2011.10.546

Vallandingham, L. R., Yu, Q., Sharma, N., Strandhagen, J. W., & Strandhagen, J. O. (2018). Grocery retail supply chain planning and control: Impact of consumer trends and enabling technologies. *IFAC-PapersOnLine*, *51*(11), 612–617. https://doi.org/10.1016/j.ifacol.2018.08.386

Vazquez-Martinez, G. A., Gonzalez-Compean, J. L., Sosa-Sosa, V. J., Morales-Sandoval, M., & Perez, J. C. (2018). CloudChain: A novel distribution model for digital products based on supply chain principles. *International Journal of Information Management*, *39*, 90–103. https://doi.org/10.1016/j.ijinfomgt.2017.12.006

Verdouw, C. N., Wolfert, J., Beulens, A. J. M., & Rialland, A. (2016). Virtualization of food supply chains with the internet of things. *Journal of Food Engineering*, *176*, 128–136. https://doi.org/10.1016/j.jfoodeng.2015.11.009

Vieira, A., Dias, L. M., Santos, M. Y., Pereira, G. A. B., & Oliveira, J. A. (2020a). On the use of simulation as a Big Data semantic validator for supply chain management. *Simulation Modelling Practice and Theory*, *98*, 101985. https://doi.org/10.1016/j.simpat.2019.101985

Vieira, A., Dias, L. M., Santos, M. Y., Pereira, G. A. B., & Oliveira, J. A. (2020b). Bypassing data issues of a supply chain simulation model in a big data context. *Procedia Manufacturing*, *42*, 132–139. https://doi.org/10.1016/j.promfg.2020.02.033

Wamba, S. F., Queiroz, M. M., & Trinchera, L. (2020). Dynamics between blockchain adoption determinants and supply chain performance: An empirical investigation. *International Journal of Production Economics*, *229*, 107791. https://doi.org/10.1016/j.ijpe.2020.107791

Wang, J., Wang, H., He, J., Li, L., Shen, M., Tan, X., Min, H., & Zheng, L. (2015). Wireless sensor network for real-time perishable food supply chain management. *Computers and Electronics in Agriculture*, *110*, 196–207. https://doi.org/10.1016/j.compag.2014.11.009

Weil, T., & Murugesan, S. (2020). IT Risk and Resilience – Cybersecurity Response to COVID-19. *IT Professional*, *22*(3), 4–10. https://doi.org/10.1109/MITP.2020.2988330

Wilson, G. (2020). *Accenture: Building Supply Chain Resilience Amidst COVID-19 | Digital Supply Chain | Supply Chain Digital*. SupplyChainDigital.Com. https://supplychaindigital.com/supply-chain-2/accenture-building-supply-chain-resilience-amidst-covid-19

Witkowski, K. (2017). Internet of things, big data, industry 4.0 – Innovative solutions in logistics and supply chains management. *Procedia Engineering*, *182*, 763–769. https://doi.org/10.1016/j.proeng.2017.03.197

Xu, Z., Elomri, A., Kerbache, L., & El Omri, A. (2020). Impacts of COVID-19 on global supply chains: Facts and perspectives. *IEEE Engineering Management Review*, *48*(3), 153–166. https://doi.org/10.1109/EMR.2020.3018420

Zhou, W., & Piramuthu, S. (2018). IoT security perspective of a flexible healthcare supply chain. *Information Technology and Management*, *19*(3), 141–153. https://doi.org/10.1007/s10799-017-0279-7

Zhu, G., Chou, M. C., & Tsai, C. W. (2020). Lessons Learned from the COVID-19 pandemic exposing the shortcomings of current supply chain operations: A long-term prescriptive offering. *Sustainability (Switzerland)*, *12*(14), 1–19. https://doi.org/10.3390/su12145858

9 Analysing the Relevance of Corporate Social Responsibility Programs in Value Chain of an Organization during COVID-19 Pandemic

Rishi Dwivedi, Smita, Ratnesh Chaturvedi, Arup Mukherjee, Amar Eron Tigga, Amanpreet Kaur, and Piyush Rai

CONTENTS

9.1 INTRODUCTION

The idea of corporate social responsibility (CSR) was first conceptualized in a comprehensive manner after the 1930s Great Depression, which created a prodigious financial, societal, and partisan ailment. CSR, also known as responsible entrepreneurship, corporate citizenship, and so on, goes beyond charity. It is one of the practical tools that unites the efforts of the corporations and social sector agencies to promote the development and sustainable growth of the entire value

DOI: 10.1201/9781003150084-9

chain of an organization, from inbound logistics to service. Using CSR, patronage, and volunteer efforts, enterprises can help society while enhancing the efficiency of the value chain. CSR can be studied from two different perspectives (Davis, 1960). Firstly, entrepreneurs identify that their responsibility to recompense to fiscal growth governs the welfare of civilization by handling the commercial element of society. Secondly, CSR is the assurance of industrialists to foster and grow social ethics: morality, teamwork, impetus, and self-fulfilment. Carroll (2008) said that CSR has transformed from merely community affairs to the corporate community and business sustainability. It also boosts the enterprise image and builds its brand. In addition, it helps in increasing employee morale and leads to greater productivity in the organization. CSR allows companies to stand out from their competitors and is the ongoing determination by businesses to give rise to economic progress while raising the quality of life of the workforce and their families (Manteaw, 2008). It overlooks profits and instead focuses on how business can satisfy all stakeholders of the valuc chain. Hoque et al. (2018) analysed that the evolution of CSR is because of commercialization and the impact that corporate houses today possess in the modern-day business environment. Moreover, enhanced education level and readily available data with respect to ecological problems and business scandals have created community doubts. Therefore, CSR is more relevant in 21st century than ever before.

In the 20th century, CSR was not made mandatory for companies. Still, the philanthropists and promoters of the company realized that they could promote the economic growth of the entire value chain only through CSR activities (Hategan et al., 2018). Besides, in the last decade of the 20th century, it can be observed that most of the developed nations have realized the importance of CSR strategies in augmenting the operational capability of enterprises and made it an important element for them (Becker-Olsen et al., 2006). The government of India made it mandatory by notification in the official gazette for companies to undertake CSR activities under the Companies Act 2013 under clause 135 (Kadyan, 2020). The organizations which have an annual turnover of Rs. 1000 crore or more, or a net profit of Rs. 5 crores, or a net worth of Rs. 500 crore or more fall under this act.

Chapple and Moon (2005) identified that the significant challenges faced while initiating CSR programs are absence of cognizance of the overall community in CSR actions, slender insight to its ingenuities, non-accessibility to efficient private partners, and so on. An enterprise consists of employees with different cultural backgrounds, traditions, and moral values. These moral values, cultural backgrounds, traditions, and so on are the factors that influence the company while choosing a CSR program, which includes education, poverty, gender equality, and hunger. Moreover, the organizations' stakeholders have a varied range of direct and hidden demands, such as superior facilities for basic amenities, sustainable resources to support the long-term necessities, and reasonable cost for products. The stakeholders of an organization can be internal as well as external. The internal stakeholders for an organization are its employees and support service partners, whereas the external stakeholders comprise customers, suppliers, and society at large. Therefore, it is imperative to apply a tool that can understand the stakeholders' requirements and suggest how those needs can suffice to achieve long-term sustainability and competence. Quality function deployment (QFD) is a model to understand customers' voices and aids in converting customers' needs and expectations into technical requirements, thereby finding the weight of each. Thus, in this chapter, the QFD model is identified as an effective technique to transform the mission and vision of the selected Indian organization into technical requirements to achieve maximum efficiency for the value chain while analysing the demands of the stakeholders of the enterprise. This chosen organization is engaged in assorted beneficence and charitable programs associated with diverse fields, such as education and livelihood, environment, healthcare, and entrepreneurship. The enduring existence and progress of the enterprise depends on attaining a subtle equilibrium between productivity and consistency with its various stakeholders through said programs.

Here, the more pertinent and demanding query is not about whether to undertake those CSR programs but about investing in education and livelihood, environment, healthcare, and entrepreneurship programs to accomplish the mutually beneficial and co-dependent social, ecological, and financial goals. Hence, there is an inherent need to evaluate the various CSR programs of the selected organization. Technique for order preference by similarity to ideal solution (TOPSIS) is one of the decision-making methods which helps rank different alternatives based on various considered criteria. The motive of this tool is to first arrive at the positive ideal solution (PIS) and a negative ideal solution (NIS), and then find an option that is closest to the perfect positive solution and far from the NIS. The PIS maximizes the profit parameters and minimizes the cost criteria, whereas the NIS maximizes the cost parameters and minimizes the benefit criteria. Then, according to the closeness to the PIS, the ranking of various alternatives is carried out. Thus, this paper applies the TOPSIS model for evaluating different CSR projects of the chosen Indian organization to satisfy the diverse needs of various stakeholders of the value chain.

9.1.1 Impact of COVID-19 on CSR Programs

The World Health Organization has declared COVID-19 a global pandemic. The infectious ailment enormously interrupted socio-economic situations of the world. Social distancing played a key part in limiting the transmission of this lethal contagion. The Indian government announced lockdowns throughout the entire country in March 2020, intending to endorse social distancing that fundamentally focused the community to keep physical and social distance. But prolonged lockdowns deteriorated fiscal predicaments. A substantial populace and an absence of cognizance further elevated the complications. Here, the significance of CSR programs becomes much more important as it can play a vital role in this pandemic situation, when individuals are trying their utmost to navigate through the turbulent phase.

Nevertheless, scenarios after the COVID-19 pandemic have exposed businesses' susceptibility to unexpected outward forces, like the 'black swan' incident of this pandemic. As the corporate settings are becoming increasingly tempestuous and unpredictable, there are a few questions which need to be answered for the long-term sustainability of the value chain of any organization. The queries which need to be answered by the managers of the enterprises are: Will organizations devote more funds in CSR actions, or will they capitulate to temporary corporate burden?

What are the programs the administrators should choose for maximizing the benefits to stakeholders? How can board members of the enterprises be persuaded of the relevance of CSR programs under growing existence fears? These problems can be responded to while critically analysing the available CSR actions available to an organization while satisfying the needs of the diverse stakeholders of its entire value chain. Therefore, an integrated QFD-TOPSIS technique is developed in this research work to analyse the demands of assorted stakeholders of an Indian organization and to make an informed decision while choosing the best CSR program, consequently augmenting the efficiency of the value chain. Additionally, the assessment of different CSR projects will help the organizations' policymakers allocate the funds more judiciously and understand the programs that are comparatively less efficient. Subsequently, corrective actions can also be carried out to enhance the efficiency of those programs.

9.2 LITERATURE REVIEW

Yan and Ma (2015) proposed an original framework to respond to uncertainties in the QFD model. Lam and Bai (2016) demonstrated that the QFD approach could enhance maritime supply chain resilience. Bolar et al. (2017) proposed an approach that utilized the hidden Markov model (HMM) for examining consumer demands through utilizing prospects of key aspects that were important to the construction industry. Wu and Liao (2018) proposed a new product planning (NPP) model to maximize customer satisfaction based on fuzzy goal programming and the QFD model. Vanany et al. (2019) developed a multi-layered QFD technique to categorize vital procedures and ranked them to augment halal food making. Dwivedi and Chakraborty (2015) proposed the application of a performance measurement tool integrating QFD and balanced scorecard (BSC) approaches to determine the growth of SME taken into consideration. Dwivedi et al. (2018) combined BSC and QFD techniques, which helped the managers understand key selling propositions for an agriculture enterprise. It further assisted the managers in devising cost-efficient strategic plans for long-term existence. Dror (2019) adopted the QFD method for prioritizing day-to-day actions of an enterprise with respect to its business goals. Pandey (2020) illustrated that the fuzzy-based QFD methodology can devise customer-centric strategic plans for an airport. Wang et al. (2020) developed a new combined quality design structure for a big multifaceted goods value chain through merging the fuzzy QFD and the grey decision-making techniques. Khan et al. (2021) applied a hybrid model while integrating a full consistency method and fuzzy QFD technique for a healthcare sector in Pakistan. Park et al. (2021) demonstrated that the QFD model could be a suitable technique in self-service technology design in cafeterias.

Gao and Hailu (2013) proposed an integrated AHP-TOPSIS technique that enhances participants' exchange of ideas and decision-making capability. Jayant et al. (2014) developed a framework to help an organization's executives in choosing and assessing diverse third-party reverse logistics providers while employing the combined AHP-TOPSIS method. Prakash and Barua (2015) proposed a methodology based on fuzzy AHP-TOPSIS technique to derive rank pre-order of the solutions of reverse logistics adoption. Sindhu et al. (2017) selected an appropriate renewable energy site in India while adopting an integrated AHP-TOPSIS model.

It can be seen from the literature review that QFD and TOPSIS methodologies have been implemented in different sectors, but an integrated QFD-TOPSIS technique has not been applied in the past for evaluating CSR programs. Hence, in this chapter, at first the QFD model is applied to evaluate important stakeholders' requirements with respect to CSR actions of a selected organization which are subsequently converted into technical requirements. Next, those technical criteria are prioritized while employing the QFD model. Further, the ranking of different CSR programs undertaken by the said enterprise is carried out based on prioritized technical requirements while using the TOPSIS technique. The results derived from the application of the proposed combined QFD-TOPSIS method can help the management of the organization to enhance the productivity and performance of CSR programs in order to achieve sustainable competitive advantage. It can also facilitate the extensive advancement of all stakeholders of the entire value chain of the selected enterprise.

9.3 QFD METHODOLOGY

QFD is a quality improvement tool developed in the late 1960s in Japan by Yoji Akao, who is regarded as the father of QFD. In 1972 under the guidance of Shigeru Mizuno and Yasushi Furukawa, this technique was first implemented at the Mitsubishi Heavy Industries Kobe Shipyard (Akao, 1990). Some of the early adopters of this technique are Toyota, Mitsubishi, IBM, Ford Motor Company, Procter and Gamble, Hewlett-Packard, Kodak, Xerox, and 3M Corporation (Prasad and Chakraborty, 2013). There are basically two components of QFD – quality and function, which are deployed in the design process of a product development. The customers' needs are brought into the product in the design phase through quality deployment. In contrast, various organizational roles and technical specifications are linked to product manufacturing by function deployment. This technique records all the said and unsaid requirements and needs of the customers, called 'voice of the customers', and then translates what the customers want to a properly designed product/service that satisfies those needs. This QFD process is done only after proper manufacturing and design planning in order to estimate a quantitative objective from a subjective measure. It also determines the relative priority of different criteria, needed for efficient service delivery. QFD technique implementation is broadly divided into four stages. First, it identifies customer requirements and gives a priority rating based on relative importance, and then those needs are converted into technical requirements and are correlated with the same to find out which is the strongest organizational need. Later, what the customer requires and how it will be done are compared and assigned a correlation index. Finally, after channelizing all the necessary efforts to meet customer requirements, the planning or design is executed, and the product or service is produced or delivered. The primary tool used in QFD for documenting the necessary information or perception about the product is known as the house of quality (HoQ) matrix. The basic structure of the HoQ matrix resembles that of a house and comprises six major building blocks, which are as follows:

- Customers' requirements comprise the customers' spoken and unspoken wants and needs in a product/service and are placed in the rows of the matrix.
- Technical requirements – These are some unique indicators that explain how the customer requirements can/may be fulfilled. They are placed in the columns of matrix.
- Technical correlation matrix – This is also known as the roof of the matrix and shows the influence of each individual technical descriptor on the other. It indicates strong, moderate, or weak relations among the technical requirements.
- Interrelationship matrix – This matrix shows the relationship between the whats and the hows while using some symbols or numbers as per each individual's contribution. In other words, it quantifies/measures the relationship between technical descriptors and customers' requirements.
- Planning matrix – The subjective measures are translated into quantifiable measures and then ranked according to their relative importance in this matrix.
- Prioritized technical requirements – They form the foundation of the HoQ matrix. Therefore, the relative importance of each technical condition is estimated in this part.

QFD systematically identifies how the end users would become interested in and satisfied with the products or services. It can be used to translate subjective quality criteria into objective ones that can be quantified and measured. The methodology is also capable of handling trade-offs between several contradicting demands of the customers. The versatility of this technique is evident from its application in various domains ranging from product design (Chen, L-H. and Ko, W-C., 2009), semiconductor industry (Chen, C-C., 2010), food processing industry (Viaene and Januszewska, 1999), non-traditional machining process (Chakraborty and Dey, 2007), software industry (Erikkson and McFadden, 1993), rapid prototyping (Ghahramani and Houshyar, 1996), construction industry (Dikmen et al., 2005), hospitality industry (Jeong and Oh, 1998), and even in the game of soccer (Partovi and Corredoira, 2002). Chan and Wu (2002) provided an extensive review on the implementation of QFD in diverse fields.

9.3.1 Development of QFD Model for Corporate Social Responsibility Program

Multiple criteria need to be analysed for making comprehensive or essential decisions. Comparing different sets of criteria can lead to ambiguity and confusion. To realize long-term existence in today's competitive environment, enterprises must evaluate their value chain concerning its operational efficiency. Appreciating the implication of this discussion, assessing enterprise's CSR performance to meet the needs of society to systematic corporate activities and intensive interaction with stakeholders seems to be of utmost importance. The data for implementing the proposed model is derived from a leading automotive components manufacturer for the original equipment manufacturers (OEMs) in India. The identity of this organization is not divulged for confidentiality purposes and subsequently will be called ABC Limited. It is incorporated under Section 8 of the Companies Act of India. ABC Limited is the CSR wing for the group. The main aim of ABC Limited is to work for community development by supporting education, livelihood promotion, women's empowerment, facilitation of healthcare, upliftment of people with disabilities, and environmental sustainability while enhancing the efficiency of the value chain of the group organizations. Thus, the customers for ABC Limited are its diverse internal and external stakeholders, such as employees, support service providers, consumers, and the society in which it is operating. The first step in the development of the QFD model for analysing factors impacting the sustenance of CSR programs consists of identifying the stakeholders' expectations while the organization performs various social actions, like planning a particular program, allocating budget to the program, and so on. Past literature review, market surveys, and clients' response forms are the channels through which the first part of QFD model is developed. The 11 most essential stakeholder requirements associated with successful completion of CSR actions of ABC Limited are identified as better skill development facility, seamless education provided to the customers, better employment opportunities, superior facilities, sustainable resources, reasonable cost, more participation of people, better working environment, empowering people, better location, and better customer redressal, which are placed along the rows of the HoQ matrix. Once the sustainability specifications of CSR programs are identified as required by the stakeholders of ABC Limited, the next step is to find the crucial technical requirements in achieving the identified stakeholders' requirements. Table 9.1 depicts the specified nine technical requirements aligned along the columns. The explanations of the selected technical requirements are discussed as follows:

1. The number of equipment pieces installed: Biometric equipment, computers, and other equipment for various CSR programs are installed in the enterprise. Those tools enhance the productivity of various CSR actions.
2. The number of personnel involved: This means the total number of staff involved in executing various activities. There should be an efficient number of personnel involved in a single program.
3. Square feet area: The square feet area is the total area in which various programs are executed. Some programs are conducted for one or two days, and other programs have a 45-day cycle.
4. The number of experts: This is different from the number of personnel. The number of personnel includes every staff member working for a particular program, but the number of experts includes only the skilled persons for a particular program. There should be a sufficient number of skilled persons so that the CSR actions can be executed properly.
5. Funds allocated (in Rs lakhs): For individual programs, a budget is prepared and the amount is allocated. This is carried out by top management, who should be prepared to look at all the expenses. This is one of the most important technical requirements to be kept in mind.

TABLE 9.1
Developed HoQ Matrix for ABC Limited

S No.	Customer requirements	ID	Priority	Number of equipment installed	Number of personnel involved	Square feet area	Number of experts	Fund allocated	Execution time	Complaint redressal time	Number of awareness programs/camps	Number of collaborations
1	Better skill development facility	1	8	9	2	7	7	5	1	4	1	8
2	Seamless education provided to the customer	1	5	1	6	1	6	4	1	1	8	2
3	Better employment opportunities	1	9	2	3	3	7	7	3	3	2	8
4	Superior facilities	1	6	5	6	6	6	6	3	2	3	4
5	Sustainable resources	1	3	1	1	1	1	8	5	2	1	4
6	Reasonable cost	-1	8	3	6	7	6	8	3	1	1	3
7	More participation of people	1	4	2	5	2	6	5	5	5	8	4
8	Better working environment	1	6	5	4	6	4	2	1	5	1	1
9	Empowering people	1	7	1	6	2	8	7	3	3	5	6
10	Better location	1	5	1	1	6	2	8	2	2	1	1
11	Better customer redressal	1	3	1	5	1	4	2	1	9	1	1
	Weightage			0.09	0.10	0.10	0.16	0.15	0.06	0.11	0.10	0.14

6. Execution time (in days): Execution time means the overall time required to implement a particular program. This should be as low as possible. If the execution time is more, it will lead to an increase in overall operational cost as well.
7. Complaint redressal time (in days): Complaint redressal time denotes the time taken to listen to and rectify an individual's problem. This signifies the efficiency of the internal business process.
8. The number of awareness programs/camps: This number clarifies the number of activities required to sensitize the villagers about the expected benefits. In other words, it is an optimal number of events needed to help villagers understand the advantages they will receive if they avail themselves of the program.
9. The number of collaborations: An organization has to collaborate with different non-governmental organizations (NGOs) and institutions to execute different CSR programs. These collaborations are significant for the enterprise in successfully implementing a particular program as it provides certain additional benefits to the villagers.

After the stakeholders' needs and technical specifications required to accomplish those needs are identified, the next step is to create the HoQ matrix. The first step is to assign values to the relationship between all pairs of stakeholder requirements of ABC Limited and technical specifications. An apposite measure of 1–9, where 1 – extremely feeble relation, 3 – feeble relation, 5 – reasonable strong relation, 7 – strong relation, and 9 – robust relation, is utilized for allocating those relationships. In the aforementioned scale, 2, 4, 6, and 8 can also be used to depict correlation as intermediate values. The recognized technical needs can either be beneficial (the higher the better) or non-beneficial (the lower the better) in nature. To represent this, nature improvement driver (ID) values are used. A beneficial criterion is depicted by a +1 value, whereas a –1 score is assigned to a non-beneficial criterion. The next stage in developing the HoQ matrix is to prioritize all the stakeholder requirements according to their relative significance. A scale of 1–5 is set, where 1 – not significant, 2 – weakly significant, 3 – significant, 4 – very significant, and 5 – extremely substantial. Once the HoQ matrix is filled up as shown in Table 9.1, the weights for each technical specification are computed utilizing the following equation:

$$W_j - ID_j \times \sum_{i=1}^{n} Pr_i \times Correlation\ Index \tag{9.1}$$

where W_j is the weight for the jth technical requirement, n is the number of customers' needs, ID_j is the value of improvement driver for the jth technical requirement, Pr_i is the priority assigned to the ith customer requirement, and correlation index is the relative importance of the jth technical requirement with respect to the ith customer requirement. The performance scores of the different CSR programs of ABC Limited are estimated in the next section while utilizing the derived weights of different technical criteria.

9.4 TOPSIS TECHNIQUE

TOPSIS is one of the most efficient techniques to solve multiple criteria problems (Hwang and Yoon, 1981). It is a more practical and straightforward methodology than non-compensatory models that include the best choice based on fixed cut-offs. The mathematical calculation required to apply this method is straightforward and unambiguous. Euclidean distance is employed to estimate the distance in the TOPSIS model, which is the basis for evaluating alternatives' performance. In this methodology, the option with minimum distance from PIS and maximum distance from NIS is identified as the best alternative. As a result, PIS maximizes the beneficial criteria and minimizes the cost criteria. In contrast, NIS selects the largest from the cost criteria and the smallest from the beneficial criteria. The following paragraphs discuss the steps involved in the TOPSIS technique.

Step 1. TOPSIS method initiates with the formation of a decision matrix. The decision matrix represents the performance measures of alternatives to various criteria.

Step 2. In the next step, different attribute dimensions are converted into non-dimensionless attributes to compare the criteria. Finally, the following expression is employed to normalize the decision matrix:

$$r_{ij} = \frac{x_{ij}}{\sqrt{\sum_{i=1}^{m} x^2_{ij}}} \tag{9.2}$$

Step 3. Subsequently, a weighted normalized decision matrix is estimated using the formulae given as follow:

$$t_{ij} = r_{ij} \Delta w_j \; i = 1, 2, \ldots, m; \; j = 1, 2, \ldots, n \tag{9.3}$$

where w_j denotes weightage of criterion C_j to the ultimate objective of the selection problem.

Step 4. The NIS and PIS are calculated employing the two following equations:

$$\begin{aligned} S_n &= \left\{ \max(t_{ij} \mid i = 1,2,\ldots.m) \mid j \in J_-, \min(t_{ij} \mid i = 1,2,\ldots.,m) \mid j \in J_+ \right\} \\ &\equiv \left\{ t_{wj} \mid j = 1,2,\ldots.n \right\} \end{aligned} \tag{9.4}$$

$$\begin{aligned} S_p &= \left\{ \min(t_{ij} \mid i = 1,2,\ldots.m) \mid j \in J_-, \max(t_{ij} \mid i = 1,2,\ldots.,m) \mid j \in J_+ \right\} \\ &\equiv \left\{ t_{bj} \mid j = 1,2,\ldots.n \right\} \end{aligned} \tag{9.5}$$

where $J_- = \{J = 1,2,\ldots.,n \mid j\}$ is associated with the criteria having a negative impact, and $J_+ = \{J = 1,2,\ldots.,n \mid j\}$ is associated with the criteria having a positive impact.

Step 5. Next, separation measure from the NIS and PIS is computed while employing the following expressions:

$$d_{iw} = \sqrt{\sum_{j=1}^{n} \left(t_{ij} - t_{wj}\right)^2} \; i = 1,2,\ldots\ldots,m \tag{9.6}$$

$$d_{ib} = \sqrt{\sum_{j=1}^{n} \left(t_{ij} - t_{bj}\right)^2} \quad i = 1,2,\ldots\ldots,m \tag{9.7}$$

Step 6. Using the following formula, estimating the relative closeness to the ideal solution is the consequent step:

$$\delta_{iw} = \frac{d_{iw}}{\left(d_{iw} + d_{ib}\right)} \quad i = 1,2,\ldots\ldots,m \tag{9.8}$$

where $0 \leq \delta_{iw} \leq 1$; $\delta_{iw} = 1$, if the alternative solution has the paramount condition, and $\delta_{iw} = 0$, if the alternative option has the worst condition.

Step 7. Finally, rank pre-order of different alternatives is derived while arranging the options in decreasing order of δ_{iw} .

9.4.1 Application of TOPSIS Approach for Ranking of CSR Programs in ABC Limited

With a view to prove the applicability, durability, and solution exactness of the integrated QFD-TOPSIS model in selecting the best CSR program for serving the diverse requirements of assorted stakeholders of ABC Limited, four major projects undertaken by it are selected: education and livelihood program, Saksham program, healthcare program, and environment program. The education and livelihood program was started by ABC Limited in 2013. This scheme facilitates superior schooling and skill development to deprived kids and teens, emphasizing the females in rural India. The Saksham program supplements the Accessible India Campaign of Government of India. It is a program that assists persons with disabilities in achieving universal accessibility. The healthcare program of ABC Limited is devoted to improving the basic healthcare infrastructure for the unprivileged society. Scrap management, water preservation, sustainable energy, tree planting, and paper saving are the actions undertaken in the environment program of ABC Limited. Next, following the steps of the considered hybrid approach, first a decision matrix is developed for the identified four CSR programs with respect to the nine chosen technical parameters, as shown in Table 9.2. The nine criteria important for achieving long-term success of CSR projects were already evaluated for their efficacy in the preceding section. After the decision matrix has been developed, it is normalized to minimize data idleness and enhance data efficiency using equation (9.2) as depicted Table 9.3. Next, Table 9.4 shows the weighted normalized matrix, which is obtained by multiplying the normalized decision matrix by the weights of diverse selected criteria using equation (9.3). Subsequently, PIS and NIS are estimated using equations (9.5) and (9.4), respectively. The separation measure from the NIS and PIS is computed in the next step based on equations (9.6) and (9.7), as shown in Table 9.5 and Table 9.6. Table 9.7 depicts the relative closeness coefficient of the said four projects of ABC Limited with respect to PIS on selected feasibility criteria along with their ranks while implementing equation (9.8). Here, the education and livelihood program is identified as the best project in ABC Limited based on stakeholders' requirements, and Saksham is the second-best choice, as exhibited in Table 9.7. It is further noted that the healthcare program is the least preferred choice.

TABLE 9.2
Decision Matrix for CSR Programs of ABC Limited

S No.	CSR program	Number of equipment installed	Number of personnel involved	Square feet area	Number of experts	Fund allocated	Execution time	Complaint redressal time	Number of awareness programs/ Camps	Number of collaborations
1	Education and livelihood program	150	29	9,007,360	18	212.92	61	2	3	4
2	Saksham program	6337	3	1500	2	37.50	44	3	23	15
3	Healthcare program	32	10	500	7	15.00	8	1	20	3
4	Environment program	20	30	600	6	172.00	12	1	10	10

TABLE 9.3
Normalized Decision Matrix for CSR Programs of ABC Limited

S No.	Corporate social responsibility program	Number of equipment installed	Number of personnel involved	Square feet area	Number of experts	Fund allocated	Execution time	Complaint redressal time	Number of awareness programs/ camps	Number of collaborations
1	Education and livelihood program	0.0236	0.6742	1.0000	0.8857	0.7695	0.7965	0.5164	0.0931	0.2138
2	Saksham program	0.9997	0.0697	0.0001	0.0984	0.1355	0.5745	0.7746	0.7139	0.8018
3	Healthcare program	0.0050	0.2324	0.0001	0.3444	0.0542	0.1044	0.2582	0.6207	0.1603
4	Environment program	0.0031	0.6974	0.0001	0.2952	0.6216	0.1567	0.2582	0.3104	0.5345

TABLE 9.4
Weighted Normalized Decision Matrix for CSR Programs of ABC Limited

S No.	CSR program	Number of equipment installed	Number of personnel involved	Square feet area	Number of experts	Fund allocated	Execution time	Complaint redressal time	Number of awareness programs/camps	Number of collaborations
1	Education and livelihood program	0.0039	0.0742	0.1100	0.0731	0.19551	0.0657	0.0102	0.0061	0.0235
2	Saksham program	0.1650	0.0077	0.0001	0.00811	0.03441	0.0474	0.0153	0.0471	0.0882
3	Healthcare program	0.0008	0.0256	0.0001	0.02841	0.01371	0.0086	0.0051	0.0409	0.0176
4	Environment program	0.0005	0.0767	0.0001	0.02431	0.1580	0.0129	0.0051	0.0205	0.0588

TABLE 9.5
Separation Measure from the NIS for CSR Programs of ABC Limited

S No.	CSR program	d_{iw}
1	Education and livelihood program	0.2223
2	Saksham program	0.1980
3	Healthcare program	0.0872
4	Environment program	0.1608

TABLE 9.6
Separation Measure from the PIS for CSR Programs of ABC Limited

S No.	CSR program	d_{ib}
1	Education and livelihood program	0.1988
2	Saksham program	0.2092
3	Healthcare program	0.2820
4	Environment program	0.2222

TABLE 9.7
Ranking of CSR Programs for ABC Limited

S No.	Corporate social responsibility program	Relative closeness coefficient	Rank
1	Education and livelihood program	0.5280	1
2	Saksham program	0.4863	2
3	Healthcare program	0.2362	4
4	Environment program	0.4197	3

9.5 RESULTS AND DISCUSSIONS

A country does not progress in the absence of financial development. Simultaneously, endorsing development is not the sole responsibility of the government. Additionally, growth becomes pointless in the absence of social inclusion. Hence, corporates play a dynamic part in societal and financial development, particularly in a country like India. The progress of an organization greatly depends on the efficiency of its value chain. CSR programs are the actions through which the diverse demands of various stakeholders of the entire value chain of the enterprise can be met and that will augment the organization's efficacy. It can be observed that COVID-19 has negatively impacted the healthiness and financial state of all sections of society. Corporates, employees, consumers, and societies have suffered a lot of loss since the outbreak of the deadly virus in 2020 in India. Therefore, it is high time that organizations should vigorously devote themselves to support diverse stakeholders in the value chain in all probable ways to achieve long-term sustainability in this perilous period. The time requires articulating competent tactical strategies and taking up varied channels, based on the enterprise's value chain, composition, and distinctiveness in association with its assorted stakeholders so that CSR programs can be best executed to meet its objectives related to financial and social growth. Here, in this research work, an integrated QFD-TOPSIS technique is applied to evaluate the stakeholders' requirements with respect to CSR programs and select the best action for ABC Limited. The results obtained by implementing the said model identify education and livelihood programs as the best project. It is in sync with ABC Limited's primary objective, which is to create more value for society. The Saksham program is chosen as the second most important action, and it promotes the organization's mission to develop shared value for all stakeholders. The information derived from the developed model can also be applied to develop a robust budget allocation according to the relative significance of the four programs. Besides, the resource allocation can also be aligned with respect to the value delivered by each CSR project. Additionally, this QFD-TOPSIS model helps administrators of ABC Limited to understand the relative significance of various parameters important for the success of the CSR projects.

9.6 CONCLUSIONS

Leading to the prospects of different business opportunities because of the globalized economy, which leads to economic growth, care should also be taken to improve an organization's value chain. Though many enterprises have adapted CSR in their operations for overall development of the value chain, it is still confusing. All organizations' projects undertaken under the name of CSR are merely philanthropy or the addition to benevolence. Various Indian companies undertake CSR activities by establishing different trusts and foundations, but these trusts and foundations are not given much significance as core business activities. This shows that Indian organizations indulge in CSR activities just because of existing tradition and not as a part of the enterprise's strategy to augment the entire value chain. To develop India as a strong nation, organizations have to agree on mandatory CSR as a part of their strategy while incorporating all parts of inbound logistics to services. Hence, this chapter demonstrates the application of an integrated QFD-TOPSIS technique for selecting the best CSR project of an Indian enterprise to fulfil various objectives of assorted stakeholders of its value chain. In addition to analysing the CSR programs based on identified technical requirements, the proposed technique provides improvement areas for projects that show huge deviation, thus helping the management of the ABC Limited improve the effectiveness of plans, policies, and strategies. The future scope of this current research work may comprise the application of this integrated QFD-TOPSIS method to evaluate CSR projects of different industries to identify the better-performing project that gives the best result. The integrated approach could easily communicate the significance of various parameters, which will also lead to attaining the long-term objectives of different CSR projects. Moreover, the organization's value chain performance can also be estimated by analysing the sustainability assessment of other CSR programs in future research work. The decisions can also be applied in solving various management problems of an organization.

REFERENCES

Akao, Y. (1990). *Quality function deployment.* Productivity Press, Cambridge.

Becker-Olsen, K. L., Cudmore, B. A., & Hill, R. P. (2006). The impact of perceived corporate social responsibility on consumer behaviour. *Journal of Business Research, 59*(1), 46–53.

Bolar, A. A., Tesfamariam, S., & Sadiq, R. (2017). Framework for prioritizing infrastructure user expectations using Quality Function Deployment (QFD). *International Journal of Sustainable Built Environment, 6*(1), 16–29.

Carroll, A. B. (2008). A history of corporate social responsibility: Concepts and practices. In *The Oxford handbook of corporate social responsibility*, 1. Oxford University Press Inc., New York.

Chakraborty, S., & Dey, S. (2007). QFD-based expert system for non-traditional machining processes selection. *Expert Systems with Applications, 32*, 1208–1217.

Chan, L-K., & Wu, M-L. (2002). Quality function deployment: A literature review. *European Journal of Operational Research, 143*, 463–497

Chapple, W., & Moon, J. (2005). Corporate social responsibility (CSR) in Asia: A seven-country study of CSR web site reporting. *Business & Society, 44*(4), 415–441.

Chen, C-C. (2010). Application of quality function deployment in the semiconductor industry: A case study. *Computers & Industrial Engineering, 58*, 672–679.

Chen, L-H., & Ko, W-C. (2009). Fuzzy approaches to quality function deployment for new product design. *Fuzzy Sets and Systems, 160*, 2620–2639.

Davis, K. (1960). Can business afford to ignore social responsibilities? *California Management Review, 2*(3), 70–76.

Dikmen, I., Birgonul, M. T., & Kiziltas, S. (2005). Strategic use of quality function deployment (QFD) in the construction industry. *Building and Environment, 40*, 245–255.

Dror, S. (2019). Linking operation plans to business objectives using QFD. *Total Quality Management & Business Excellence, 30*(1–2), 135–150.

Dwivedi, R., & Chakraborty, S. (2015). Strategy formulation and monitoring of a SME using activity-based costing, balanced scorecard, and quality function deployment models. *Transformations in Business & Economics, 14*(1), 173–191.

Dwivedi, R., Chakraborty, S., Sinha, A. K., Singh, S., & Richa. (2018). *Development of a performance measurement tool for an agricultural enterprise using BSC and QFD models.* International Conference on Mechanical, Materials and Renewable Energy. SMIT, Rangpo, Sikkim, India: IOP Conference Series: Materials Science and Engineering.

Erikkson, I., & McFadden, F. (1993). Quality function deployment: A tool to improve software quality. *Information and Software Technology, 35*, 491–498.

Gao, L., & Hailu, A. (2013). Identifying preferred management options: An integrated agent-based recreational fishing simulation model with an AHP-TOPSIS evaluation method. *Ecological Modelling, 249*, 75–83.

Ghahramani, B., & Houshyar, A. (1996). Benchmarking the application of quality function deployment in rapid prototyping. *Journal of Materials Processing Technology, 61*, 201–206.

Hategan, C. D., Sirghi, N., Curea-Pitorac, R. I., & Hategan, V. P. (2018). Doing well or doing good: The relationship between corporate social responsibility and profit in Romanian companies. *Sustainability, 10*(4), 1041.

Hoque, N., Rahman, A. R. A., Molla, R. I., Noman, A. H. M., & Bhuiyan, M. Z. H. (2018). Is corporate social responsibility pursuing pristine business goals for sustainable development?. *Corporate Social Responsibility and Environmental Management, 25*(6), 1130–1142.

Hwang, C. L., & Yoon, K. (1981). Methods for multiple attribute decision making. In *Multiple attribute decision making* (pp. 58–191). Springer, Berlin, Heidelberg.

Jayant, A., Gupta, P., Garg, S. K., & Khan, M. (2014). TOPSIS-AHP based approach for selection of reverse logistics service provider: A case study of mobile phone industry. *Procedia Engineering, 97*, 2147–2156.

Jeong, M., & Oh, H. (1998). Quality function deployment: An extended framework for service quality and customer satisfaction in the hospitality industry. *International Journal of Hospitality Management, 17*, 375–390.

Kadyan, J. S. (2020). Corporate social responsibility in India. In *The Palgrave handbook of corporate social responsibility* (pp. 1–31). Palgrave Macmillan, Cham.

Khan, F., Ali, Y., & Pamucar, D. (2021). A new fuzzy FUCOM-QFD approach for evaluating strategies to enhance the resilience of the healthcare sector to combat the COVID-19 pandemic. *Kybernetes, ahead-of-print*(ahead-of-print).

Lam, J. S. L., & Bai, X. (2016). A quality function deployment approach to improve maritime supply chain resilience. *Transportation Research Part E: Logistics and Transportation Review, 92*, 16–27.

Manteaw, B. (2008). From tokenism to social justice. Rethinking the bottom line for sustainable community development. *Community Development Journal, 43*(4), 428–443.

Pandey, M. M. (2020). Evaluating the strategic design parameters of airports in Thailand to meet service expectations of Low-Cost Airlines using the Fuzzy-based QFD method. *Journal of Air Transport Management, 82*, 101738.

Park, S., Lehto, X., & Lehto, M. (2021). Self-service technology kiosk design for restaurants: An QFD application. *International Journal of Hospitality Management, 92*, 102757.

Partovi, F. Y., & Corredoira, R. A. (2002). Quality function deployment for the good of soccer. *European Journal of Operational Research, 137*, 642–656.

Prakash, C., & Barua, M. K. (2015). Integration of AHP-TOPSIS method for prioritizing the solutions of reverse logistics adoption to overcome its barriers under fuzzy environment. *Journal of Manufacturing Systems, 37*, 599–615.

Prasad, K., & Chakraborty, S. (2013). A quality function deployment-based model for materials selection. *Materials and Design, 49*, 525–535.

Sindhu, S., Nehra, V., & Luthra, S. (2017). Investigation of feasibility study of solar farms deployment using hybrid AHP-TOPSIS analysis: Case study of India. *Renewable and Sustainable Energy Reviews, 73*, 496–511.

Vanany, I., Maarif, G. A., & Soon, J. M. (2019). Application of multi-based Quality Function Deployment (QFD) model to improve halal meat industry. *Journal of Islamic Marketing, 10*(1), 97–124.

Viaene, J., & Januszewska, R. (1999). Quality function deployment in the chocolate industry. *Food Quality and Preference, 10*, 377–385.

Wang, H., Fang, Z., Wang, D., & Liu, S. (2020). An integrated fuzzy QFD and grey decision-making approach for supply chain collaborative quality design of large complex products. *Computers & Industrial Engineering, 140*, 106212.

Wu, X., & Liao, H. (2018). An approach to quality function deployment based on probabilistic linguistic term sets and ORESTE method for multi-expert multi-criteria decision making. *Information Fusion, 43*, 13–26.

Yan, H. B., & Ma, T. (2015). A group decision-making approach to uncertain quality function deployment based on fuzzy preference relation and fuzzy majority. *European Journal of Operational Research, 241*(3), 815–829.

Index